AF663841

BOOLEAN ALGEBRA

BOOLEAN ALGEBRA

BY
HARWELL WEBSTER
ALISAH TURNER

KNOWLEDGE BAKERS

AUTHORS

)78-93-9001-360-9 (HB)

l rights reserved under International Copyright Conventions. No part of t thereof, including title of the book, be reproduced in any printed form or la : written permission of the author and the publishers.

ed by : **Knowledge Bakers**
Pune, India
Email: info@knowledgebakers.com

PREFACE

the realm of digital logic and computer science, Boolean algebra se
rock upon which modern computing systems are built. Originating fr
ing work of George Boole in the 19th century, Boolean algebra pro
s framework for representing and manipulating logical expressions, e
ign and analysis of complex digital circuits and algorithms. As we del
icacies of Boolean algebra, it becomes apparent that its principles ext
the confines of electronics, permeating various fields such as mathe
phy, and engineering.

is book, aptly titled "Boolean Algebra," aims to provide a compre
tion of this fundamental mathematical discipline. Through collaborati
and scholars in the field, we endeavor to elucidate the theoretical underpi
l methodologies, and diverse applications of Boolean algebra in both ac
ustry.

editors of this volume, we have curated a collection of chapters tha
ntial concepts and techniques of Boolean algebra, including logic gate
Boolean functions, minimization methods, and more. Each chapter offe
tions, illustrative examples, and practical exercises to aid readers in g
icacies of Boolean algebra and applying its principles to solve rea
ns.

thermore, this book seeks to highlight the broad spectrum of appli
olean algebra encompasses, ranging from digital circuit design and co
ture to database systems, artificial intelligence, and beyond. By showcasi
interdisciplinary connections, and emerging trends, we hope to dem
satility and relevance of Boolean algebra in contemporary scienti
ogical endeavors.

we embark on this enlightening journey through the realms of E
, we extend our gratitude to the contributors whose expertise and dec
ıriched this volume. We hope that "Boolean Algebra" serves as a v
e for students, researchers, and practitioners alike, inspiring curiosity, fc
anding, and fostering innovation in this foundational field of mathemat
:er science.

Harwell V
Alisah

CONTENTS

1

Boolean Algebra: Overview

In mathematics and mathematical logic, Boolean algebra is the branch of algebra in which the values of the variables are the truth values true and false, usually denoted 1 and 0, respectively. Instead of elementary algebra, where the values of the variables are numbers and the prime operations are addition and multiplication, the main operations of Boolean algebra are the conjunction (and) denoted as '", the disjunction (or) denoted as (", and the negation (not) denoted as ¬. It is thus a formalism for describing logical operations, in the same way that elementary algebra describes numerical operations.

Boolean algebra was introduced by George Boole in his first book The Mathematical Analysis of Logic (1847), and set forth more fully in his An Investigation of the Laws of Thought (1854). According to Huntington, the term "Boolean algebra" was first suggested by Sheffer in 1913, although Charles Sanders Peirce gave the title "A Boolean Algebra with One Constant" to the first chapter of his "The Simplest Mathematics" in 1880. Boolean algebra has been fundamental in the development of digital electronics, and is provided for in all modern programming languages. It is also used in set theory and statistics.

Boolean algebra is the algebra of two-valued logic with only sentential connectives, or equivalently of algebras of sets under union and complementation. The rigorous concept is that of a certain kind of algebra, analogous to the mathematical notion of a group. This concept has roots and applications in logic (Lindenbaum-Tarski algebras and model theory), set theory (fields of sets), topology (totally disconnected compact Hausdorff spaces), foundations of set theory (Boolean-valued models), measure theory (measure algebras), functional analysis (algebras of projections), and ring theory (Boolean rings).

The study of Boolean algebras has several aspects: structure theory, model theory of Boolean algebras, decidability and undecidability questions for the class of Boolean algebras, and the indicated applications. In addition, although not explained here, there are connections to other logics, subsumption as a part of special kinds of algebraic logic, finite Boolean algebras and switching circuit theory, and Boolean matrices.

The distinguishing factor of Boolean algebra is that it deals only with the study of binary variables. Most commonly Boolean variables are presented with the possible values of 1 ("true") or 0 ("false"). Variables can also have more complex interpretations, such as in set theory. Boolean algebra is also known as binary algebra.

DEFINITION AND SIMPLE PROPERTIES

A Boolean algebra (BA) is a set A together with binary operations + and . and a unary operation ", and elements 0, 1 of A such that the following laws hold: commutative and associative laws for addition and multiplication, distributive laws both for multiplication over addition and for addition over multiplication, and the following special laws:

$x+(x.y) = x$

$x.(x+y) = x$

$x+(\wedge x) = 1$

$x.(\wedge x) = 0$

These laws are better understood in terms of the basic example of a BA, consisting of a collection A of subsets of a set X closed under the operations of union, intersection, complementation with respect to X, with members " and X. One can easily derive many elementary laws from these axioms, keeping in mind this example for motivation. Any BA has a natural partial order ≤ defined upon it by saying that $x \leq y$ if and only if $x+y = y$. This corresponds in our main example to Í. Of special importance is the two-element BA, formed by taking the set X to have just one element. The two-element BA shows the direct connection with elementary logic. The two members, 0 and 1, correspond to falsity and truth, respectively. The Boolean operations then express the ordinary truth tables for disjunction (with +), conjunction (with .) and negation (with –).

An important elementary result is that an equation holds in all BAs if and only if it holds in the two-element BA. Next, we define $x \oplus y=(x.-y)+(y.-x)$. Then A together with ⊕ and ., along with 0 and 1, forms a ring with identity in which every element is idempotent. Conversely, given such a ring, with addition ⊕ and multiplication, define $x+y=x \oplus y \oplus (x.y)$ and "$x=1 \oplus x$. This makes the ring into a BA. These two processes are inverses of one another, and show that the theory of Boolean algebras and of rings with identity in which every element is idempotent are definitionally equivalent. This puts the theory of BAs into a standard object of research in algebra. An atom in a BA is a nonzero element such that there is no element b with $0<b<a$. A BA is atomic if every nonzero element of the BA is above an atom. Finite BAs are atomic, but so are many infinite BAs.

Under the partial order $\leq$ above, x+y is the least upper bound of x and y, and x.y is the greatest lower bound of x and y. We can generalize this: ÓX is the least upper bound of a set X of elements, and ĐX is the greatest lower bound of a set X of elements. These do not exist for all sets in all Boolean algebras; if they do always exist, the Boolean algebra is said to be complete.

History of Boolean Algebra

A precursor of Boolean algebra was Gottfried Wilhelm Leibniz's algebra of concepts. Leibniz's algebra of concepts is deductively equivalent to the Boolean algebra of sets. Boole's algebra predated the modern developments in abstract algebra and mathematical logic; it is however seen as connected to the origins of both fields. In an abstract setting, Boolean algebra was perfected in the late 19th century by Jevons, Schröder, Huntington and others, until it reached the modern conception of an (abstract) mathematical structure. For example, the empirical observation that one can manipulate expressions in the algebra of sets, by translating them into expressions in Boole's algebra, is explained in modern terms by saying that the algebra of sets is a Boolean algebra (note the indefinite article). In fact, M. H. Stone proved in 1936 that every Boolean algebra is isomorphic to a field of sets.

In the 1930s, while studying switching circuits, Claude Shannon observed that one could also apply the rules of Boole's algebra in this setting, and he introduced switching algebra as a way to analyze and design circuits by algebraic means in terms of logic gates. Shannon already had at his disposal the abstract mathematical apparatus, thus he cast his switching algebra as the two-element Boolean algebra. In modern circuit engineering settings, there is little need to consider other Boolean algebras, thus "switching algebra" and "Boolean algebra" are often used interchangeably.

Efficient implementation of Boolean functions is a fundamental problem in the design of combinational logic circuits. Modern electronic design automation tools for VLSI circuits often rely on an efficient representation of Boolean functions known as (reduced ordered) binary decision diagrams (BDD) for logic synthesis and formal verification.

Logic sentences that can be expressed in classical propositional calculus have an equivalent expression in Boolean algebra. Thus, Boolean logic is sometimes used to denote propositional calculus performed in this way.

Boolean algebra is not sufficient to capture logic formulas using quantifiers, like those from first order logic.

Although the development of mathematical logic did not follow Boole's program, the connection between his algebra and logic was later put on firm ground in the setting of algebraic logic, which also studies the algebraic systems of many other logics. The problem of determining whether the variables of a given Boolean (propositional) formula can be assigned in such a way as to make the formula evaluate to true is called the Boolean satisfiability problem (SAT), and is of importance to theoretical computer science, being the first problem shown to be NP-complete. The closely related model of computation known as a Boolean circuit relates time complexity (of an algorithm) to circuit complexity.

UNDERSTANDING BOOLEAN ALGEBRA

Boolean algebra is different from elementary algebra as the latter deals with numerical operations and the former deals with logical operations. Elementary algebra is expressed using basic mathematical functions, such as addition, subtraction, multiplication, and division, whereas Boolean algebra deals with conjunction, disjunction, and negation.

The concept of Boolean algebra was first introduced by George Boole in his book "The Mathematical Analysis of Logic," and further expanded upon in his book "An Investigation of the Laws of Thought." Since its concept has been detailed, Boolean algebra's primary use has been in computer programming languages. Its mathematical purposes are used in set theory and statistics.

The elementary algebraic theory

Several algebraic constructions have obvious definitions and simple properties for BAs: subalgebras, homomorphisms, isomorphisms, and direct products (even of infinitely many algebras). Some other standard algebraic constructions are more peculiar to BAs. An ideal in a BA is a subset I closed under +, with 0 as a member, and such that if a ≤ b"I, then also a"I. Although not immediately obvious, this is the same as the ring-theoretic concept. There is a dual notion of a filter (with no counterpart in rings in general). A filter is a subset F closed under . , having 1 as a member, and such that if ae"b"F, then also a"F. An ultrafilter on A is a filter F with the following properties: 0 "F, and for any a"A, either a"F or "a"F. For any a"A, let S(a) = {F:F is an ultrafilter on A and a"F}.

Then S is an isomorphism onto a BA of subsets of the set X of all ultrafilters on A. This establishes the basic Stone representation theorem, and clarifies the origin of BAs as concrete algebras of sets. Moreover, the sets S(a) can be declared to be a base for a topology on X, and this turns X into a totally disconnected compact Hausdorff space. This establishes a one-one correspondence between the class of BAs and the class of such spaces. As a consequence, used very much in the theory of BAs, many topological theorems and concepts have consequences for BAs. If x is an element of a BA, we let 0x=″x and 1x=x. If (x(0), …x(m″1)) is a finite sequence of elements of a BA A, then every element of the subalgebra of A generated by {x(0),…,x(m″1)} can be written as a sum of monomials e(0)x(0).....e(m″1)x(m″1) for e in some set of functions mapping m={0,…,m″1} into 2={0,1}.

This is an algebraic expression of the disjunctive normal form theorem of sentential logic. A function f from a set X of generators of a BA A into a BA B can be extended to a homomorphism if and only if e(0)x(0).....e(m″1)x(m″1)=0 always implies that e (0) f (x (0)).....e(m″1)f(x(m″1))=0.

This is Sikorski's extension criterion. Every BA A can be embedded in a complete BA B in such a way that every element of B is the least upper bound of a set of elements of A. B is unique up to A-isomorphism, and is called the completion of A. If f is a homomorphism from a BA A into a complete BA B, and if A is a subalgebra of C, then f can be extended to a homomorphism of C into B. This is Sikorski's extension theorem. Another general algebraic notion which applies to Boolean algebras is the notion of a free algebra. This can be concretely constructed for BAs. Namely, the free BA on ê is the BA of closed-open subsets of the two element discrete space raised to the ê power.

Special classes of Boolean algebras

There are many special classes of Boolean algebra which are important both for the intrinsic theory of BAs and for applications:

- Atomic BAs, already mentioned above.
- Atomless BAs, which are defined to be BAs without any atoms. For example, any infinite free BA is atomless.
- Complete BAs, defined above. These are especially important in the foundations of set theory.
- Interval algebras. These are derived from linearly ordered sets (L,<) with a first element as follows. One takes the smallest algebra of subsets of L containing all of the half-open intervals [a,b) with a in L and b in L or equal to ″. These BAs are useful in the study of Lindenbaum-Tarski algebras. Every countable BA is isomorphic to an interval algebra, and thus a countable BA can be described by indicating an ordered set such that it is isomorphic to the corresponding interval algebra.

Tree algebras. A tree is a partially ordered set (T,<) in which the set of predecessors of any element is well-ordered. Given such a tree, one considers the algebra of subsets of T generated by all sets of the form {b:a≤b} for some a in T.

Superatomic BAs. These are BAs which are not only atomic, but are such that each subalgebra and homomorphic image is atomic.

Values

Whereas expressions denote mainly numbers in elementary algebra, in Boolean algebra, they denote the truth values false and true. These values are represented with the bits (or binary digits), namely 0 and 1. They do not behave like the integers 0 and 1, for which 1 + 1 = 2, but may be identified with the elements of the two-element field GF(2), that is, integer arithmetic modulo 2, for which 1 + 1 = 0. Addition and multiplication then play the Boolean roles of XOR (exclusive-or) and AND (conjunction), respectively, with disjunction x (“ y (inclusive-or) definable as x + y - xy and negation ¬x as 1 “ x. In GF(2), “ may be replaced by +, since they denote the same operation; however this way of writing Boolean operations allows applying the usual arithmetic operations of integers (this may be useful when using a programming language in which GF(2) is not implemented).

Boolean algebra also deals with functions which have their values in the set {0, 1}. A sequence of bits is a commonly used for such functions. Another common example is the subsets of a set E: to a subset F of E, one can define the indicator function that takes the value 1 on F, and 0 outside F. The most general example is the elements of a Boolean algebra, with all of the foregoing being instances thereof.

As with elementary algebra, the purely equational part of the theory may be developed, without considering explicit values for the variables.

Structure theory and cardinal functions on Boolean algebras

Much of the deeper theory of Boolean algebras, telling about their structure and classification, can be formulated in terms of certain functions defined for all Boolean algebras, with infinite cardinals as values. We define some of the more important of these cardinal functions, and state some of the known structural facts, mostly formulated in terms of them:

1. The cellularity c(A) of a BA is the supremum of the cardinalities of sets of pairwise disjoint elements of A.
2. A subset X of a BA A is independent if X is a set of free generators of the subalgebra that it generates. The independence of A is the supremum of cardinalities of independent subsets of A.

3. A subset X of a BA A is dense in A if every nonzero element of A is e″ a nonzero element of X. The ð-weight of A is the smallest cardinality of a dense subset of A.
4. Two elements x,y of A are incomparable if neither one is ≤ the other. The supremum of cardinalities of subset X of A consisting of pairwise incomparable elements is the incomparability of A.
5. A subset X of A is irredundant if no element of X is in the subalgebra generated by the others.

An important fact concerning cellularity is the Erdõs-Tarski theorem: if the cellularity of a BA is a singular cardinal, then there actually is a set of disjoint elements of that size; for cellularity regular limit (inaccessible), there are counterexamples. Every infinite complete BA has an independent subset of the same size as the algebra. Every infinite BA A has an irredundant incomparable subset whose size is the ð-weight of A.

Every interval algebra has countable independence. A superatomic algebra does not even have an infinite independent subset. Every tree algebra can be embedded in an interval algebra. A BA with only the identity automorphism is called rigid. There exist rigid complete BAs, also rigid interval algebras and rigid tree algebras.

More recently, many cardinal functions of min-max type have been studied. For example, small independence is the smallest size of an infinite maximal independent set; and small cellularity is the smallest size of an infinite partition of unity.

DECIDABILITY AND UNDECIDABILITY QUESTIONS

A basic result of Tarski is that the elementary theory of Boolean algebras is decidable. Even the theory of Boolean algebras with a distinguished ideal is decidable. On the other hand, the theory of a Boolean algebra with a distinguished subalgebra is undecidable. Both the decidability results and undecidablity results extend in various ways to Boolean algebras in extensions of first-order logic.

Lindenbaum-Tarski algebras

A very important construction, which carries over to many logics and many algebras other than Boolean algebras, is the construction of a Boolean algebra associated with the sentences in some logic. The simplest case is sentential logic. Here there are sentence symbols, and common connectives building up longer sentences from them: disjunction, conjunction, and negation. Given a set A of sentences in this language, two sentences s and t are equivalent modulo A if and only if the biconditional between them is a logical consequence of A.

The equivalence classes can be made into a BA such that + corresponds to disjunction, . to conjunction, and " to negation. Any BA is isomorphic to one of this form. One can do something similar for a first-order theory. Let T be a first-order theory in a first-order language L. We call formulas Õ and ø equivalent provided that T¢"Õ"!ø. The equivalence class of a sentence Õ is denoted by [Õ]. Let A be the collection of all equivalence classes under this equivalence relation. We can make A into a BA by the following definitions, which are easily justified:

$$[\phi]+[\psi] = [\phi \vee \psi]$$
$$[\phi]\cdot[\psi] = [\phi \wedge \psi]$$
$$-[\phi] = [\neg\phi]$$
$$0 = [F]$$
$$1 = [T]$$

Every BA is isomorphic to a Lindenbaum-Tarski algebra. However, one of the most important uses of these classical Lindenbaum-Tarski algebras is to describe them for important theories (usually decidable theories). For countable languages this can be done by describing their isomorphic interval algebras. Generally this gives a thorough knowledge of the theory.

Boolean-valued models

In model theory, one can take values in any complete BA rather than the two-element BA. This Boolean-valued model theory was developed around 1950–1970, but has not been worked on much since. But a special case, Boolean-valued models for set theory, is very much at the forefront of current research in set theory. It actually forms an equivalent way of looking at the forcing construction of Cohen, and has some technical advantages and disadvantages. Philosophically it seems more satisfactory than the forcing concept. We describe this set theory case here; it will then become evident why only complete BAs are considered. Let B be a complete BA. First we define the Boolean valued universe V(B). The ordinary set-theoretic universe can be identified with V(2), where 2 is the 2-element BA. The definition is by transfinite recursion, where á,â are ordinals and ë is a limit ordinal:

$$V(B,0) = \varnothing$$
$$V(B,\alpha+1) = \{f : \mathrm{dom}(f) \subset V(B,\alpha) \text{and range}(f) \subset B\}$$
$$V(B,\lambda) = \bigcup_{\beta<\lambda} V(B,\beta).$$

where dom(f) is the domain of function f and range(f) is the range of function f. The B-valued universe is the proper class V(B) which is the union of all of these Vs. Next, one defines by a rather complicated transfinite recursion over well-founded sets the value of a set-theoretic formula with elements of the Boolean valued universe assigned to its free variables:

$$\begin{aligned}
\| x \in y \| &= \sum_{t \in dom(y)} \| x = t \| \cdot y(t) \\
\| x \subseteq y \| &= \prod_{t \in dom(x)} - x(t) + \| t \in y \| \\
\| x = y \| &= \| x \subseteq y \| \cdot \| y \subseteq x \| \\
\| \neg \phi \| &= - \| \phi \| \\
\| \phi \vee \psi \| &= \| \phi \| + \| \psi \| \\
\| \exists x \phi(x) \| &= \sum_{a \in V(B)} \| \phi(a) \| .
\end{aligned}$$

Boolean Algebra in Finance

Boolean algebra has applications in finance through mathematical modeling of market activities. For example, research into the pricing of stock options can be aided by the use of a binary tree to represent the range of possible outcomes in the underlying security. In this binomial options pricing model, where there are only two possible outcomes, the Boolean variable represents an increase or a decrease in the price of the security.

This type of modeling is necessary because, in American options, which can be exercised at any time, the path of a security's price is just as important as its final price. The binomial options pricing model requires the path of a security's price to be broken into a series of discrete time ranges.

As such, the binomial options pricing model allows an investor or trader to view the change in the asset price from one period to the next. This allows them to evaluate the option based on decisions made at different points.

Because a U.S. based option can be exercised at any time, this allows a trader to determine whether they should exercise an option or hold onto it for a longer period. An analysis of the binomial tree would allow a trader to see in advance if an option should be exercised. If there is a positive value, then the option should be exercised, if the value is negative, then the trader should hold onto the position.

Laws

A law of Boolean algebra is an identity such as x (" (y (" z) = (x (" y) (" z between two Boolean terms, where a Boolean term is defined as an expression built up from variables and the constants 0 and 1 using the operations '", (", and ¬. The concept can be extended to terms involving other Boolean operations such as ⊕, →, and a″, but such extensions are unnecessary for the purposes to which the laws are put. Such purposes include the definition of a Boolean algebra as any model of the Boolean laws, and as a means for deriving new laws from old as in the derivation of x (" (y ‴ z) = x (" (z ‴ y) from y ‴ z = z ‴ y (as treated in § Axiomatizing Boolean algebra).

Laws of Boolean Algebra

There are six types of Boolean algebra laws. They are:

- Commutative law
- Associative law
- Distributive law
- AND law
- OR law
- Inversion law

Those six laws are explained in detail here.

Commutative Law

Any binary operation which satisfies the following expression is referred to as a commutative operation. Commutative law states that changing the sequence of the variables does not have any effect on the output of a logic circuit.

- $A. B = B. A$
- $A + B = B + A$

Associative Law

It states that the order in which the logic operations are performed is irrelevant as their effect is the same.

- $(A. B). C = A . (B . C)$
- $(A + B) + C = A + (B + C)$

Distributive Law

Distributive law states the following conditions:

- $A. (B + C) = (A. B) + (A. C)$
- $A + (B. C) = (A + B) . (A + C)$

AND Law

These laws use the AND operation. Therefore they are called AND laws.

- $A .0 = 0$
- $A . 1 = A$
- $A. A = A$

- $A.\bar{A} = 0$

OR Law

These laws use the OR operation. Therefore they are called OR laws.

- $A + 0 = A$
- $A + 1 = 1$
- $A + A = A$
- $A + \bar{A} = 1$

Inversion Law

In Boolean algebra, the inversion law states that double inversion of variable results in the original variable itself.

COMPLETENESS

The laws listed above define Boolean algebra, in the sense that they entail the rest of the subject. The laws Complementation 1 and 2, together with the monotone laws, suffice for this purpose and can therefore be taken as one possible complete set of laws or axiomatization of Boolean algebra. Every law of Boolean algebra follows logically from these axioms. Furthermore, Boolean algebras can then be defined as the models of these axioms as treated in § Boolean algebras.

To clarify, writing down further laws of Boolean algebra cannot give rise to any new consequences of these axioms, nor can it rule out any model of them. In contrast, in a list of some but not all of the same laws, there could have been Boolean laws that did not follow from those on the list, and moreover there would have been models of the listed laws that were not Boolean algebras.

This axiomatization is by no means the only one, or even necessarily the most natural given that we did not pay attention to whether some of the axioms followed from others but simply chose to stop when we noticed we had enough laws, treated further in § Axiomatizing Boolean algebra. Or the intermediate notion of axiom can be sidestepped altogether by defining a Boolean law directly as any tautology, understood as an equation that holds for all values of its variables over 0 and 1. All these definitions of Boolean algebra can be shown to be equivalent.

Duality principle

Principle: If {X, R} is a poset, then {X, R(inverse)} is also a poset.

There is nothing magical about the choice of symbols for the values of Boolean algebra. We could rename 0 and 1 to say á and â, and as long as we did so consistently throughout it would still be Boolean algebra, albeit with some obvious cosmetic differences.

But suppose we rename 0 and 1 to 1 and 0 respectively. Then it would still be Boolean algebra, and moreover operating on the same values. However it would not be identical to our original Boolean algebra because now we find (∨ behaving the way ∧ used to do and vice versa. So there are still some cosmetic differences to show that we've been fiddling with the notation, despite the fact that we're still using 0s and 1s.

But if in addition to interchanging the names of the values we also interchange the names of the two binary operations, now there is no trace of what we have done. The end product is completely indistinguishable from what we started with. We might notice that the columns for x ∨y and x (∨y in the truth tables had changed places, but that switch is immaterial.

When values and operations can be paired up in a way that leaves everything important unchanged when all pairs are switched simultaneously, we call the members of each pair dual to each other. Thus 0 and 1 are dual, and ∨and (∨are dual. The Duality Principle, also called De Morgan duality, asserts that Boolean algebra is unchanged when all dual pairs are interchanged.

One change we did not need to make as part of this interchange was to complement. We say that complement is a self-dual operation. The identity or do-nothing operation x (copy the input to the output) is also self-dual. A more complicated example of a self-dual operation is (x∨y) (∨(y∨z) (∨(z ∨x). There is no self-dual binary operation that depends on both its arguments. A composition of self-dual operations is a self-dual operation. For example, if f(x, y, z) = (x∨y) (∨(y ∨z) (∨(z ∨x), then f(f(x, y, z), x, t) is a self-dual operation of four arguments x, y, z, t.

The principle of duality can be explained from a group theory perspective by the fact that there are exactly four functions that are one-to-one mappings (automorphisms) of the set of Boolean polynomials back to itself: the identity function, the complement function, the dual function and the contradual function (complemented dual). These four functions form a group under function composition, isomorphic to the Klein four-group, acting on the set of Boolean polynomials. Walter Gottschalk remarked that consequently a more appropriate name for the phenomenon would be the principle (or square) of quaternality.

Boolean Algebra Theorems

The two important theorems which are extremely used in Boolean algebra are De Morgan's First law and De Morgan's second law. These two theorems are used to change the Boolean expression. This theorem basically helps to reduce the given Boolean expression in the simplified form. These two De Morgan's laws are used to change the expression from one form to another form. Now, let us discuss these two theorems in detail.

De Morgan's First Law

De Morgan's First Law states that (A.B)′ = A′+B′.

The first law states that the complement of the product of the variables is equal to the sum of their individual complements of a variable.

The truth table that shows the verification of De Morgan's First law is given as follows:

A	B	A'	B'	(A.B)'	A'+B'
0	0	1	1	1	1
0	1	1	0	1	1
1	0	0	1	1	1
1	1	0	0	0	0

The last two columns show that (A.B)′ = A′+B′.

Hence, De Morgan's First Law is proved.

De Morgan's Second Law:

De Morgan's Second law states that (A+B)′ = A′. B′.

The second law states that the complement of the sum of variables is equal to the product of their individual complements of a variable.

The following truth table shows the proof for De Morgan's second law.

A	B	A'	B'	(A+B)'	A'. B'
0	0	1	1	1	1
0	1	1	0	0	0
1	0	0	1	0	0
1	1	0	0	0	0

The last two columns show that (A+B)′ = A′. B′.

Hence, De Morgan's second law is proved.

The other theorems in Boolean algebra are complementary theorem, duality theorem, transposition theorem, redundancy theorem and so on. All these theorems are used to simplify the given Boolean expression. The reduced Boolean expression should be equivalent to the given Boolean expression.

Diagrammatic representations

Venn diagrams

A Venn diagram can be used as a representation of a Boolean operation using shaded overlapping regions. There is one region for each variable, all circular in the examples here. The interior and exterior of region x corresponds respectively to the values 1 (true) and 0 (false) for variable x. The shading indicates the value of the operation for each combination of regions, with dark denoting 1 and light 0 (some authors use the opposite convention).

The three Venn diagrams in the figure below represent respectively conjunction $x \wedge y$, disjunction x(“y, and complement $\neg x$.

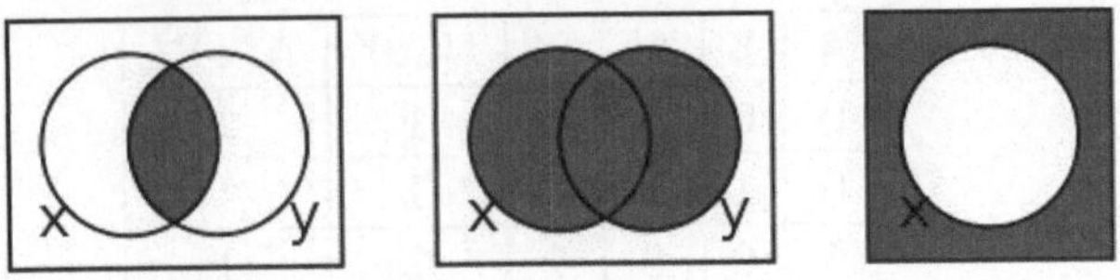

$x \wedge y$ $x \vee y$ $\neg x$

Venn diagrams for conjunction, disjunction, and complement

For conjunction, the region inside both circles is shaded to indicate that $x \wedge y$ is 1 when both variables are 1. The other regions are left unshaded to indicate that $x \wedge y$ is 0 for the other three combinations.

The second diagram represents disjunction x(“y by shading those regions that lie inside either or both circles. The third diagram represents complement $\neg x$ by shading the region not inside the circle.

While we have not shown the Venn diagrams for the constants 0 and 1, they are trivial, being respectively a white box and a dark box, neither one containing a circle. However we could put a circle for x in those boxes, in which case each would denote a function of one argument, x, which returns the same value independently of x, called a constant function. As far as their outputs are concerned, constants and constant functions are indistinguishable; the difference is that a constant takes no arguments, called a zeroary or nullary operation, while a constant function takes one argument, which it ignores, and is a unary operation.

Venn diagrams are helpful in visualizing laws. The commutativity laws for ‘‘’ and (“ can be seen from the symmetry of the diagrams: a binary operation that was not commutative would not have a symmetric diagram because interchanging x and y would have the effect of reflecting the diagram horizontally and any failure of commutativity would then appear as a failure of symmetry.

Idempotence of ‘‘’ and (“ can be visualized by sliding the two circles together and noting that the shaded area then becomes the whole circle, for both ‘‘’ and.

To see the first absorption law, x∧(x(“y) = x, start with the diagram in the middle for x(“y and note that the portion of the shaded area in common with the x circle is the whole of the x circle. For the second absorption law, x(“(x∧y) = x, start with the left diagram for x∧y and note that shading the whole of the x circle results in just the x circle being shaded, since the previous shading was inside the x circle.

The double negation law can be seen by complementing the shading in the third diagram for ¬x, which shades the x circle.

To visualize the first De Morgan’s law, (¬x)∧(¬y) = ¬(x(“y), start with the middle diagram for x(“y and complement its shading so that only the region outside both circles is shaded, which is what the right hand side of the law describes. The result is the same as if we shaded that region which is both outside the x circle and outside the y circle, i.e. the conjunction of their exteriors, which is what the left hand side of the law describes.

The second De Morgan’s law, (¬x)(“(¬y) = ¬(x∧y), works the same way with the two diagrams interchanged.

The first complement law, x∧¬x = 0, says that the interior and exterior of the x circle have no overlap. The second complement law, x(“¬x = 1, says that everything is either inside or outside the x circle.

Digital logic gates

Digital logic is the application of the Boolean algebra of 0 and 1 to electronic hardware consisting of logic gates connected to form a circuit diagram. Each gate implements a Boolean operation, and is depicted schematically by a shape indicating the operation. The shapes associated with the gates for conjunction (AND-gates), disjunction (OR-gates), and complement (inverters) are as follows.

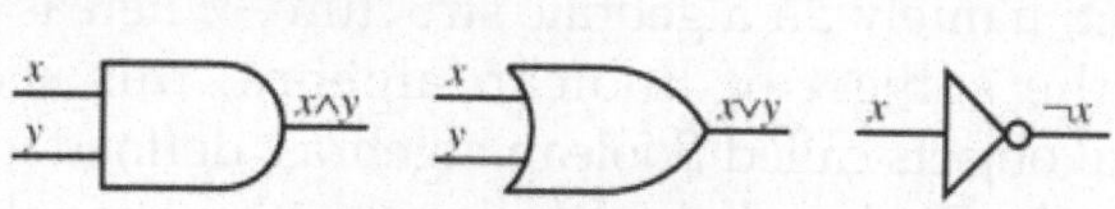

Figure 3. Logic gates

The lines on the left of each gate represent input wires or ports. The value of the input is represented by a voltage on the lead. For so-called “active-high” logic, 0 is represented by a voltage close to zero or “groun≤, while 1 is represented by a voltage close to the supply voltage; active-low reverses this. The line on the right of each gate represents the output port, which normally follows the same voltage conventions as the input ports.

Complement is implemented with an inverter gate. The triangle denotes the operation that simply copies the input to the output; the small circle on the output denotes the actual inversion complementing the input. The convention of putting such a circle on any port means that the signal passing through this port is complemented on the way through, whether it is an input or output port. The Duality Principle, or De Morgan’s laws, can be understood as asserting that complementing all three ports of an AND gate converts it to an OR gate and vice versa, as shown in Figure 4 below. Complementing both ports of an inverter however leaves the operation unchanged.

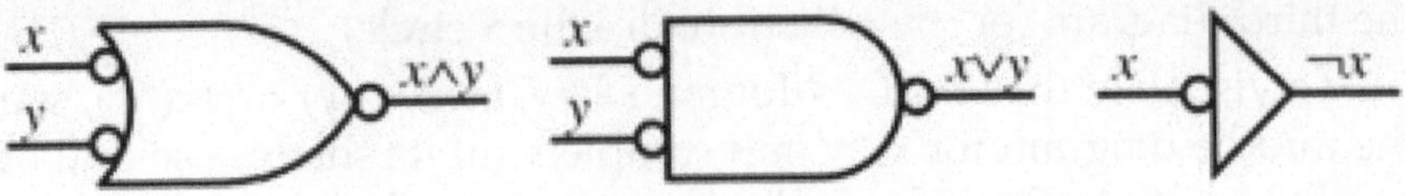

Figure 4. De Morgan equivalents

More generally one may complement any of the eight subsets of the three ports of either an AND or OR gate. The resulting sixteen possibilities give rise to only eight Boolean operations, namely those with an odd number of 1's in their truth table. There are eight such because the "odd-bit-out" can be either 0 or 1 and can go in any of four positions in the truth table. There being sixteen binary Boolean operations, this must leave eight operations with an even number of 1's in their truth tables. Two of these are the constants 0 and 1 (as binary operations that ignore both their inputs); four are the operations that depend nontrivially on exactly one of their two inputs, namely x, y, $\neg x$, and $\neg y$; and the remaining two are $x \oplus y$ (XOR) and its complement xa"y.

BOOLEAN ALGEBRAS STRUCTURE

The term "algebra" denotes both a subject, namely the subject of algebra, and an object, namely an algebraic structure. Whereas the foregoing has addressed the subject of Boolean algebra, this section deals with mathematical objects called Boolean algebras, defined in full generality as any model of the Boolean laws. We begin with a special case of the notion definable without reference to the laws, namely concrete Boolean algebras, and then give the formal definition of the general notion.

Concrete Boolean algebras

A concrete Boolean algebra or field of sets is any nonempty set of subsets of a given set X closed under the set operations of union, intersection, and complement relative to X.

(As an aside, historically X itself was required to be nonempty as well to exclude the degenerate or one-element Boolean algebra, which is the one exception to the rule that all Boolean algebras satisfy the same equations since the degenerate algebra satisfies every equation. However this exclusion conflicts with the preferred purely equational definition of "Boolean algebra", there being no way to rule out the one-element algebra using only equations— 0 `" 1 does not count, being a negated equation.

Hence modern authors allow the degenerate Boolean algebra and let X be empty.)

Example 1. The power set 2^X of X, consisting of all subsets of X. Here X may be any set: empty, finite, infinite, or even uncountable.

Example 2. The empty set and X. This two-element algebra shows that a concrete Boolean algebra can be finite even when it consists of subsets of an infinite set. It can be seen that every field of subsets of X must contain the empty set and X. Hence no smaller example is possible, other than the degenerate algebra obtained by taking X to be empty so as to make the empty set and X coincide.

Example 3. The set of finite and cofinite sets of integers, where a cofinite set is one omitting only finitely many integers. This is clearly closed under complement, and is closed under union because the union of a cofinite set with any set is cofinite, while the union of two finite sets is finite. Intersection behaves like union with "finite" and "cofinite" interchanged.

Example 4. For a less trivial example of the point made by Example 2, consider a Venn diagram formed by n closed curves partitioning the diagram into 2^n regions, and let X be the (infinite) set of all points in the plane not on any curve but somewhere within the diagram. The interior of each region is thus an infinite subset of X, and every point in X is in exactly one region. Then the set of all 2^{2n} possible unions of regions (including the empty set obtained as the union of the empty set of regions and X obtained as the union of all 2^n regions) is closed under union, intersection, and complement relative to X and therefore forms a concrete Boolean algebra. Again we have finitely many subsets of an infinite set forming a concrete Boolean algebra, with Example 2 arising as the case n = 0 of no curves.

Subsets as bit vectors

A subset Y of X can be identified with an indexed family of bits with index set X, with the bit indexed by x " X being 1 or 0 according to whether or not x " Y. (This is the so-called characteristic function notion of a subset.) For example, a 32-bit computer word consists of 32 bits indexed by the set {0,1,2,...,31}, with 0 and 31 indexing the low and high order bits respectively. For a smaller example, if X = {a,b,c} where a, b, c are viewed as bit positions in that order from left to right, the eight subsets {}, {c}, {b}, {b,c}, {a}, {a,c}, {a,b}, and {a,b,c} of X can be identified with the respective bit vectors 000, 001, 010, 011, 100, 101, 110, and 111. Bit vectors indexed by the set of natural numbers are infinite sequences of bits, while those indexed by the reals in the unit interval [0,1] are packed too densely to be able to write conventionally but nonetheless form well-defined indexed families (imagine coloring every point of the interval [0,1] either black or white independently;

the black points then form an arbitrary subset of [0,1]). From this bit vector viewpoint, a concrete Boolean algebra can be defined equivalently as a nonempty set of bit vectors all of the same length (more generally, indexed by the same set) and closed under the bit vector operations of bitwise '", (", and ¬, as in 1010'"0110 = 0010, 1010("0110 = 1110, and ¬1010 = 0101, the bit vector realizations of intersection, union, and complement respectively.

THE PROTOTYPICAL BOOLEAN ALGEBRA

The set {0,1} and its Boolean operations as treated above can be understood as the special case of bit vectors of length one, which by the identification of bit vectors with subsets can also be understood as the two subsets of a one-element set. We call this the prototypical Boolean algebra, justified by the following observation.

The laws satisfied by all nondegenerate concrete Boolean algebras coincide with those satisfied by the prototypical Boolean algebra.

This observation is easily proved as follows. Certainly any law satisfied by all concrete Boolean algebras is satisfied by the prototypical one since it is concrete. Conversely any law that fails for some concrete Boolean algebra must have failed at a particular bit position, in which case that position by itself furnishes a one-bit counterexample to that law. Nondegeneracy ensures the existence of at least one bit position because there is only one empty bit vector.

The final goal of the next section can be understood as eliminating "concrete" from the above observation. We shall however reach that goal via the surprisingly stronger observation that, up to isomorphism, all Boolean algebras are concrete.

Boolean algebras: the definition

The Boolean algebras we have seen so far have all been concrete, consisting of bit vectors or equivalently of subsets of some set. Such a Boolean algebra consists of a set and operations on that set which can be shown to satisfy the laws of Boolean algebra.

Instead of showing that the Boolean laws are satisfied, we can instead postulate a set X, two binary operations on X, and one unary operation, and require that those operations satisfy the laws of Boolean algebra. The elements of X need not be bit vectors or subsets but can be anything at all. This leads to the more general abstract definition.

A Boolean algebra is any set with binary operations '" and (" and a unary operation ¬ thereon satisfying the Boolean laws.

For the purposes of this definition it is irrelevant how the operations came to satisfy the laws, whether by fiat or proof. All concrete Boolean algebras satisfy the laws (by proof rather than fiat), whence every concrete Boolean algebra is a Boolean algebra according to our definitions. This axiomatic definition of a Boolean algebra as a set and certain operations satisfying certain laws or axioms by fiat is entirely analogous to the abstract definitions of group, ring, field etc. characteristic of modern or abstract algebra.

Given any complete axiomatization of Boolean algebra, such as the axioms for a complemented distributive lattice, a sufficient condition for an algebraic structure of this kind to satisfy all the Boolean laws is that it satisfy just those axioms. The following is therefore an equivalent definition.

A Boolean algebra is a complemented distributive lattice.

The section on axiomatization lists other axiomatizations, any of which can be made the basis of an equivalent definition.

Representable Boolean algebras

Although every concrete Boolean algebra is a Boolean algebra, not every Boolean algebra need be concrete. Let n be a square-free positive integer, one not divisible by the square of an integer, for example 30 but not 12. The operations of greatest common divisor, least common multiple, and division into n (that is, $\neg x = n/x$), can be shown to satisfy all the Boolean laws when their arguments range over the positive divisors of n. Hence those divisors form a Boolean algebra. These divisors are not subsets of a set, making the divisors of n a Boolean algebra that is not concrete according to our definitions.

However, if we represent each divisor of n by the set of its prime factors, we find that this nonconcrete Boolean algebra is isomorphic to the concrete Boolean algebra consisting of all sets of prime factors of n, with union corresponding to least common multiple, intersection to greatest common divisor, and complement to division into n. So this example while not technically concrete is at least "morally" concrete via this representation, called an isomorphism. This example is an instance of the following notion.

A Boolean algebra is called representable when it is isomorphic to a concrete Boolean algebra.

The obvious next question is answered positively as follows.

Every Boolean algebra is representable.

That is, up to isomorphism, abstract and concrete Boolean algebras are the same thing. This quite nontrivial result depends on the Boolean prime ideal theorem, a choice principle slightly weaker than the axiom of choice, and is treated in more detail in the article Stone's representation theorem for Boolean algebras. This strong relationship implies a weaker result strengthening the observation in the previous subsection to the following easy consequence of representability.

The laws satisfied by all Boolean algebras coincide with those satisfied by the prototypical Boolean algebra.

It is weaker in the sense that it does not of itself imply representability. Boolean algebras are special here, for example a relation algebra is a Boolean algebra with additional structure but it is not the case that every relation algebra is representable in the sense appropriate to relation algebras.

BOOLEAN ALGEBRA (STRUCTURE)

Definition

A Boolean algebra is a six-tuple consisting of a set A, equipped with two binary operations '" (called "meet" or "an≤), (" (called "join" or "or"), a unary operation ¬ (called "complement" or "not") and two elements 0 and 1 in A (called "bottom" and "top", or "least" and "greatest" element, also denoted by the symbols ¥" and £", respectively), such that for all elements a, b and c of A, the following axioms hold:

$a \vee (b \vee c) = (a \vee b) \vee c$	$a \wedge (b \wedge c) = (a \wedge b) \wedge c$	associativity
$a \vee b = b \vee a$	$a \wedge b = b \wedge a$	commutativity
$a \vee (a \wedge b) = a$	$a \wedge (a \vee b) = a$	absorption
$a \vee 0 = a$	$a \wedge 1 = a$	identity
$a \vee (b \wedge c) = (a \vee b) \wedge (a \vee c)$	$a \wedge (b \vee c) = (a \wedge b) \vee (a \wedge c)$	distributivity
$a \vee \neg a = 1$	$a \wedge \neg a = 0$	complements

Note, however, that the absorption law and even the associativity law can be excluded from the set of axioms as they can be derived from the other axioms (see Proven properties).

A Boolean algebra with only one element is called a trivial Boolean algebra or a degenerate Boolean algebra. (In older works, some authors required 0 and 1 to be distinct elements in order to exclude this case.)[citation needed]

It follows from the last three pairs of axioms above (identity, distributivity and complements), or from the absorption axiom, that

$$a = b \wedge a \quad \text{if and only if} \quad a \vee b = b.$$

The relation d∨defined by a d∨b if these equivalent conditions hold, is a partial order with least element 0 and greatest element 1. The meet a∨b and the join a (∨b of two elements coincide with their infimum and supremum, respectively, with respect to d∨ The first four pairs of axioms constitute a definition of a bounded lattice.

It follows from the first five pairs of axioms that any complement is unique.

The set of axioms is self-dual in the sense that if one exchanges (∨with ∨and 0 with 1 in an axiom, the result is again an axiom. Therefore, by applying this operation to a Boolean algebra (or Boolean lattice), one obtains another Boolean algebra with the same elements; it is called its dual.

Examples

- The simplest non-trivial Boolean algebra, the two-element Boolean algebra, has only two elements, 0 and 1, and is defined by the rules:

∧ 0 1			∨ 0 1			*a* 0 1
0 0 0			0 0 1			¬*a* 1 0
1 0 1			1 1 1			

- It has applications in logic, interpreting 0 as false, 1 as true, ''' as and, ('' as or, and ¬ as not. Expressions involving variables and the Boolean operations represent statement forms, and two such expressions can be shown to be equal using the above axioms if and only if the corresponding statement forms are logically equivalent.
- The two-element Boolean algebra is also used for circuit design in electrical engineering;[note 1] here 0 and 1 represent the two different states of one bit in a digital circuit, typically high and low voltage. Circuits are described by expressions containing variables, and two such expressions are equal for all values of the variables if and only if the corresponding circuits have the same input-output behavior. Furthermore, every possible input-output behavior can be modeled by a suitable Boolean expression.
- The two-element Boolean algebra is also important in the general theory of Boolean algebras, because an equation involving several variables is generally true in all Boolean algebras if and only if it is true in the two-element Boolean algebra (which can be checked by a trivial brute force algorithm for small numbers of variables). This can for example be used to show that the following laws (Consensus theorems) are generally valid in all Boolean algebras:
  - o (a (∨ b) ∧ (¬a (∧c) ∧ (b (∨ c) a≡ (a (∨ b) ∧ (¬a (∨c)
  - o (a ∧ b) (∨ (¬a ∧ c) (∨ (b ∧ c) a≡ (a ∨ b) (∨ (¬a ∨ c)

- The power set (set of all subsets) of any given nonempty set S forms a Boolean algebra, an algebra of sets, with the two operations (" := *" (union) and '" :=)" (intersection). The smallest element 0 is the empty set and the largest element 1 is the set S itself.
- After the two-element Boolean algebra, the simplest Boolean algebra is that defined by the power set of two atoms:

	∧ **0 a b 1**			∨ **0 a b 1**			*x* **0 a b 1**
	0 0 0 0 0			**0** 0 a b 1			¬*x* 1 b a 0
	a 0 a 0 a			**a** a a 1 1			
	b 0 0 b b			**b** b 1 b 1			
	1 0 a b 1			**1** 1 1 1 1			

- The set A of all subsets of S that are either finite or cofinite is a Boolean algebra and an algebra of sets called the finite–cofinite algebra. If S is infinite then the set of all cofinite subsets of S, which is called the Fréchet filter, is a free ultrafilter on A. However, the Fréchet filter is not an ultrafilter on the power set of S.
- Starting with the propositional calculus with ê sentence symbols, form the Lindenbaum algebra (that is, the set of sentences in the propositional calculus modulo logical equivalence). This construction yields a Boolean algebra. It is in fact the free Boolean algebra on ê generators. A truth assignment in propositional calculus is then a Boolean algebra homomorphism from this algebra to the two-element Boolean algebra.
- Given any linearly ordered set L with a least element, the interval algebra is the smallest algebra of subsets of L containing all of the half-open intervals [a, b) such that a is in L and b is either in L or equal to ". Interval algebras are useful in the study of Lindenbaum–Tarski algebras; every countable Boolean algebra is isomorphic to an interval algebra.
  Homomorphisms and isomorphisms

  A homomorphism between two Boolean algebras A and B is a function

$f : A \to B$ such that for all a, b in A:

f(a (∨ b) = f(a) (∨ f(b),

$f(a \wedge b) = f(a) \wedge f(b)$,

$f(0) = 0$,

$f(1) = 1$.

It then follows that f(¬a) = ¬f(a) for all a in A. The class of all Boolean algebras, together with this notion of morphism, forms a full subcategory of the category of lattices.

An isomorphism between two Boolean algebras A and B is a homomorphism f : A → B with an inverse homomorphism, that is, a homomorphism g : B → A such that the composition g ‘“ f: A → A is the identity function on A, and the composition f ‘“ g: B → B is the identity function on B. A homomorphism of Boolean algebras is an isomorphism if and only if it is bijective.

BOOLEAN RINGS

Every Boolean algebra (A, ‘“, (“) gives rise to a ring (A, +, ·) by defining a + b := (a ‘“ ¬b) (“ (b ‘“ ¬a) = (a (“ b) ‘“ ¬(a ‘“ b) (this operation is called symmetric difference in the case of sets and XOR in the case of logic) and a · b := a ‘“ b. The zero element of this ring coincides with the 0 of the Boolean algebra; the multiplicative identity element of the ring is the 1 of the Boolean algebra. This ring has the property that a · a = a for all a in A; rings with this property are called Boolean rings.

Conversely, if a Boolean ring A is given, we can turn it into a Boolean algebra by defining x (“ y := x + y + (x · y) and x ‘“ y := x · y. Since these two constructions are inverses of each other, we can say that every Boolean ring arises from a Boolean algebra, and vice versa. Furthermore, a map f : A → B is a homomorphism of Boolean algebras if and only if it is a homomorphism of Boolean rings. The categories of Boolean rings and Boolean algebras are equivalent.

Hsiang (1985) gave a rule-based algorithm to check whether two arbitrary expressions denote the same value in every Boolean ring. More generally, Boudet, Jouannaud, and Schmidt-Schauß (1989) gave an algorithm to solve equations between arbitrary Boolean-ring expressions. Employing the similarity of Boolean rings and Boolean algebras, both algorithms have applications in automated theorem proving.

Ideals and filters

An ideal of the Boolean algebra A is a subset I such that for all x, y in I we have x (“ y in I and for all a in A we have a ‘“ x in I. This notion of ideal coincides with the notion of ring ideal in the Boolean ring A. An ideal I of A is called prime if I `” A and if a ‘“ b in I always implies a in I or b in I. Furthermore, for every a “ A we have that a ‘“ -a = 0 “ I and then a “ I or -a “ I for every a “ A, if I is prime. An ideal I of A is called maximal if I `” A and if the only ideal properly containing I is A itself. For an ideal I, if a “ I and -a “ I, then I *” {a} or I *” {-a} is properly contained in another ideal J.

Hence, that an I is not maximal and therefore the notions of prime ideal and maximal ideal are equivalent in Boolean algebras. Moreover, these notions coincide with ring theoretic ones of prime ideal and maximal ideal in the Boolean ring A.

The dual of an ideal is a filter. A filter of the Boolean algebra A is a subset p such that for all x, y in p we have x "' y in p and for all a in A we have a (" x in p. The dual of a maximal (or prime) ideal in a Boolean algebra is ultrafilter. Ultrafilters can alternatively be described as 2-valued morphisms from A to the two-element Boolean algebra. The statement every filter in a Boolean algebra can be extended to an ultrafilter is called the Ultrafilter Theorem and cannot be proven in ZF, if ZF is consistent. Within ZF, it is strictly weaker than the axiom of choice. The Ultrafilter Theorem has many equivalent formulations: every Boolean algebra has an ultrafilter, every ideal in a Boolean algebra can be extended to a prime ideal, etc.

REPRESENTATIONS

It can be shown that every finite Boolean algebra is isomorphic to the Boolean algebra of all subsets of a finite set. Therefore, the number of elements of every finite Boolean algebra is a power of two.

Stone's celebrated representation theorem for Boolean algebras states that every Boolean algebra A is isomorphic to the Boolean algebra of all clopen sets in some (compact totally disconnected Hausdorff) topological space.

Axiomatics

The first axiomatization of Boolean lattices/algebras in general was given by the English philosopher and mathematician Alfred North Whitehead in 1898. It included the above axioms and additionally x("1=1 and $x \wedge 0 = 0$. In 1904, the American mathematician Edward V. Huntington (1874–1952) gave probably the most parsimonious axiomatization based on "', (", ¬, even proving the associativity laws (see box). He also proved that these axioms are independent of each other. In 1933, Huntington set out the following elegant axiomatization for Boolean algebra. It requires just one binary operation + and a unary functional symbol n, to be read as 'complement', which satisfy the following laws:

1. Commutativity: $x + y = y + x$.
2. Associativity: $(x + y) + z = x + (y + z)$.
3. Huntington equation: $n(n(x) + y) + n(n(x) + n(y)) = x$.

Herbert Robbins immediately asked: If the Huntington equation is replaced with its dual, to wit:

4. Robbins Equation: $n(n(x + y) + n(x + n(y))) = x$,

do (1), (2), and (4) form a basis for Boolean algebra? Calling (1), (2), and (4) a Robbins algebra, the question then becomes: Is every Robbins algebra a Boolean algebra? This question (which came to be known as the Robbins conjecture) remained open for decades, and became a favorite question of Alfred Tarski and his students. In 1996, William McCune at Argonne National Laboratory, building on earlier work by Larry Wos, Steve Winker, and Bob Veroff, answered Robbins's question in the affirmative: Every Robbins algebra is a Boolean algebra. Crucial to McCune's proof was the computer program EQP he designed. For a simplification of McCune's proof, see Dahn (1998).

Further work has been done for reducing the number of axioms; see Minimal axioms for Boolean algebra.

GENERALIZATIONS

Removing the requirement of existence of a unit from the axioms of Boolean algebra yields "generalized Boolean algebras". Formally, a distributive lattice B is a generalized Boolean lattice, if it has a smallest element 0 and for any elements a and b in B such that $a \leq b$, there exists an element x such that a '" x = 0 and a (" x = b. Defining a " b as the unique x such that (a '" b) (" x = a and (a '" b) '" x = 0, we say that the structure (B,∧,(",",0) is a generalized Boolean algebra, while (B,(",0) is a generalized Boolean semilattice. Generalized Boolean lattices are exactly the ideals of Boolean lattices.

A structure that satisfies all axioms for Boolean algebras except the two distributivity axioms is called an orthocomplemented lattice. Orthocomplemented lattices arise naturally in quantum logic as lattices of closed subspaces for separable Hilbert spaces.

Axiomatizing Boolean algebra

The above definition of an abstract Boolean algebra as a set and operations satisfying "the" Boolean laws raises the question, what are those laws? A simple-minded answer is "all Boolean laws", which can be defined as all equations that hold for the Boolean algebra of 0 and 1. Since there are infinitely many such laws this is not a terribly satisfactory answer in practice, leading to the next question: does it suffice to require only finitely many laws to hold?

In the case of Boolean algebras the answer is yes. In particular the finitely many equations we have listed above suffice. We say that Boolean algebra is finitely axiomatizable or finitely based.

Can this list be made shorter yet? Again the answer is yes. To begin with, some of the above laws are implied by some of the others. A sufficient subset of the above laws consists of the pairs of associativity, commutativity, and absorption laws, distributivity of '" over (" (or the other distributivity law—one suffices), and the two complement laws. In fact this is the traditional axiomatization of Boolean algebra as a complemented distributive lattice.

By introducing additional laws not listed above it becomes possible to shorten the list yet further; for instance, with the vertical bar representing the Sheffer stroke operation, the single axiom ((a #" b) #" c) #" (a #" ((a #" c) #" a)) = c is sufficient to completely axiomatize Boolean algebra. It is also possible to find longer single axioms using more conventional operations; see Minimal axioms for Boolean algebra.

Propositional logic

Propositional logic is a logical system that is intimately connected to Boolean algebra. Many syntactic concepts of Boolean algebra carry over to propositional logic with only minor changes in notation and terminology, while the semantics of propositional logic are defined via Boolean algebras in a way that the tautologies (theorems) of propositional logic correspond to equational theorems of Boolean algebra.

Syntactically, every Boolean term corresponds to a propositional formula of propositional logic. In this translation between Boolean algebra and propositional logic, Boolean variables x,y... become propositional variables (or atoms) P,Q,..., Boolean terms such as x("y become propositional formulas P("Q, 0 becomes false or ¥", and 1 becomes true or T. It is convenient when referring to generic propositions to use Greek letters Ö, Ø,... as metavariables (variables outside the language of propositional calculus, used when talking about propositional calculus) to denote propositions.

The semantics of propositional logic rely on truth assignments. The essential idea of a truth assignment is that the propositional variables are mapped to elements of a fixed Boolean algebra, and then the truth value of a propositional formula using these letters is the element of the Boolean algebra that is obtained by computing the value of the Boolean term corresponding to the formula. In classical semantics, only the two-element Boolean algebra is used, while in Boolean-valued semantics arbitrary Boolean algebras are considered. A tautology is a propositional formula that is assigned truth value 1 by every truth assignment of its propositional variables to an arbitrary Boolean algebra (or, equivalently, every truth assignment to the two element Boolean algebra).

These semantics permit a translation between tautologies of propositional logic and equational theorems of Boolean algebra. Every tautology Ö of propositional logic can be expressed as the Boolean equation Ö = 1, which will be a theorem of Boolean algebra. Conversely every theorem Ö = Ø of Boolean algebra corresponds to the tautologies (Ö("¬Ø) ‴ (¬Ö("Ø) and (Ö∧Ø) (" (¬Ö∧¬Ø). If → is in the language these last tautologies can also be written as (Ö→Ø) ‴ (Ø→Ö), or as two separate theorems Ö→Ø and Ø→Ö; if a" is available then the single tautology Ö a" Ø can be used.

Applications

One motivating application of propositional calculus is the analysis of propositions and deductive arguments in natural language. Whereas the proposition "if x = 3 then x+1 = 4" depends on the meanings of such symbols as + and 1, the proposition "if x = 3 then x = 3" does not; it is true merely by virtue of its structure, and remains true whether "x = 3" is replaced by "x = 4" or "the moon is made of green cheese." The generic or abstract form of this tautology is "if P then P", or in the language of Boolean algebra, "P → P".

Replacing P by x = 3 or any other proposition is called instantiation of P by that proposition. The result of instantiating P in an abstract proposition is called an instance of the proposition. Thus "x = 3 → x = 3" is a tautology by virtue of being an instance of the abstract tautology "P → P". All occurrences of the instantiated variable must be instantiated with the same proposition, to avoid such nonsense as P → x = 3 or x = 3 → x = 4.

Propositional calculus restricts attention to abstract propositions, those built up from propositional variables using Boolean operations. Instantiation is still possible within propositional calculus, but only by instantiating propositional variables by abstract propositions, such as instantiating Q by Q→P in P→(Q→P) to yield the instance P→((Q→P)→P).

(The availability of instantiation as part of the machinery of propositional calculus avoids the need for metavariables within the language of propositional calculus, since ordinary propositional variables can be considered within the language to denote arbitrary propositions. The metavariables themselves are outside the reach of instantiation, not being part of the language of propositional calculus but rather part of the same language for talking about it that this sentence is written in, where we need to be able to distinguish propositional variables and their instantiations as being distinct syntactic entities.)

DEDUCTIVE SYSTEMS FOR PROPOSITIONAL LOGIC

An axiomatization of propositional calculus is a set of tautologies called axioms and one or more inference rules for producing new tautologies from old. A proof in an axiom system A is a finite nonempty sequence of propositions each of which is either an instance of an axiom of A or follows by some rule of A from propositions appearing earlier in the proof (thereby disallowing circular reasoning). The last proposition is the theorem proved by the proof. Every nonempty initial segment of a proof is itself a proof, whence every proposition in a proof is itself a theorem. An axiomatization is sound when every theorem is a tautology, and complete when every tautology is a theorem.

Sequent calculus

Propositional calculus is commonly organized as a Hilbert system, whose operations are just those of Boolean algebra and whose theorems are Boolean tautologies, those Boolean terms equal to the Boolean constant 1. Another form is sequent calculus, which has two sorts, propositions as in ordinary propositional calculus, and pairs of lists of propositions called sequents, such as A("B, A∧C,... ¢" A, B→C,.... The two halves of a sequent are called the antecedent and the succedent respectively. The customary metavariable denoting an antecedent or part thereof is Ã, and for a succedent ∇; thus Ã, A ¢" ∇ would denote a sequent whose succedent is a list ∇ and whose antecedent is a list Ã with an additional proposition A appended after it. The antecedent is interpreted as the conjunction of its propositions, the succedent as the disjunction of its propositions, and the sequent itself as the entailment of the succedent by the antecedent.

Entailment differs from implication in that whereas the latter is a binary operation that returns a value in a Boolean algebra, the former is a binary relation which either holds or does not hold. In this sense entailment is an external form of implication, meaning external to the Boolean algebra, thinking of the reader of the sequent as also being external and interpreting and comparing antecedents and succedents in some Boolean algebra. The natural interpretation of ¢" is as ≤ in the partial order of the Boolean algebra defined by x ≤ y just when x("y = y. This ability to mix external implication ¢" and internal implication → in the one logic is among the essential differences between sequent calculus and propositional calculus.

Applications

Boolean algebra as the calculus of two values is fundamental to computer circuits, computer programming, and mathematical logic, and is also used in other areas of mathematics such as set theory and statistics.

Computers

In the early 20th century, several electrical engineers intuitively recognized that Boolean algebra was analogous to the behavior of certain types of electrical circuits. Claude Shannon formally proved such behavior was logically equivalent to Boolean algebra in his 1937 master's thesis, A

Symbolic Analysis of Relay and Switching Circuits.

Today, all modern general purpose computers perform their functions using two-value Boolean logic; that is, their electrical circuits are a physical manifestation of two-value Boolean logic. They achieve this in various ways: as voltages on wires in high-speed circuits and capacitive storage devices, as orientations of a magnetic domain in ferromagnetic storage devices, as holes in punched cards or paper tape, and so on. (Some early computers used decimal circuits or mechanisms instead of two-valued logic circuits.)

Of course, it is possible to code more than two symbols in any given medium. For example, one might use respectively 0, 1, 2, and 3 volts to code a four-symbol alphabet on a wire, or holes of different sizes in a punched card. In practice, the tight constraints of high speed, small size, and low power combine to make noise a major factor. This makes it hard to distinguish between symbols when there are several possible symbols that could occur at a single site. Rather than attempting to distinguish between four voltages on one wire, digital designers have settled on two voltages per wire, high and low.

Computers use two-value Boolean circuits for the above reasons. The most common computer architectures use ordered sequences of Boolean values, called bits, of 32 or 64 values, e.g. 01101000110101100101010101001011. When programming in machine code, assembly language, and certain other programming languages, programmers work with the low-level digital structure of the data registers. These registers operate on voltages, where zero volts represents Boolean 0, and a reference voltage (often +5 V, +3.3 V, +1.8 V) represents Boolean 1. Such languages support both numeric operations and logical operations. In this context, "numeric" means that the computer treats sequences of bits as binary numbers (base two numbers) and executes arithmetic operations like add, subtract, multiply, or divide. "Logical" refers to the Boolean logical operations of disjunction, conjunction, and negation between two sequences of bits, in which each bit in one sequence is simply compared to its counterpart in the other sequence. Programmers therefore have the option of working in and applying the rules of either numeric algebra or Boolean algebra as needed. A core differentiating feature between these families of operations is the existence of the carry operation in the first but not the second.

Two-valued logic

Other areas where two values is a good choice are the law and mathematics. In everyday relaxed conversation, nuanced or complex answers such as "maybe" or "only on the weeken≤ are acceptable. In more focused situations such as a court of law or theorem-based mathematics however it is deemed advantageous to frame questions so as to admit a simple yes-or-no answer—is the defendant guilty or not guilty, is the proposition true or false—and to disallow any other answer. However much of a straitjacket this might prove in practice for the respondent, the principle of the simple yes-no question has become a central feature of both judicial and mathematical logic, making two-valued logic deserving of organization and study in its own right.

A central concept of set theory is membership. Now an organization may permit multiple degrees of membership, such as novice, associate, and full. With sets however an element is either in or out. The candidates for membership in a set work just like the wires in a digital computer: each candidate is either a member or a nonmember, just as each wire is either high or low.

Algebra being a fundamental tool in any area amenable to mathematical treatment, these considerations combine to make the algebra of two values of fundamental importance to computer hardware, mathematical logic, and set theory.

Two-valued logic can be extended to multi-valued logic, notably by replacing the Boolean domain {0, 1} with the unit interval [0,1], in which case rather than only taking values 0 or 1, any value between and including 0 and 1 can be assumed. Algebraically, negation (NOT) is replaced with 1 ″ x, conjunction (AND) is replaced with multiplication (xy), and disjunction (OR) is defined via De Morgan's law. Interpreting these values as logical truth values yields a multi-valued logic, which forms the basis for fuzzy logic and probabilistic logic. In these interpretations, a value is interpreted as the "degree" of truth – to what extent a proposition is true, or the probability that the proposition is true.

BOOLEAN OPERATIONS

The original application for Boolean operations was mathematical logic, where it combines the truth values, true or false, of individual formulas.

Natural language

Natural languages such as English have words for several Boolean operations, in particular conjunction (and), disjunction (or), negation (not), and implication (implies). But not is synonymous with and not. When used to combine situational assertions such as "the block is on the table" and "cats drink milk," which naively are either true or false, the meanings of these logical connectives often have the meaning of their logical counterparts. However, with descriptions of behavior such as "Jim walked through the door", one starts to notice differences such as failure of commutativity, for example the conjunction of "Jim opened the door" with "Jim walked through the door" in that order is not equivalent to their conjunction in the other order, since and usually means and then in such cases.

Questions can be similar: the order "Is the sky blue, and why is the sky blue?" makes more sense than the reverse order. Conjunctive commands about behavior are like behavioral assertions, as in get dressed and go to school. Disjunctive commands such love me or leave me or fish or cut bait tend to be asymmetric via the implication that one alternative is less preferable. Conjoined nouns such as tea and milk generally describe aggregation as with set union while tea or milk is a choice. However context can reverse these senses, as in your choices are coffee and tea which usually means the same as your choices are coffee or tea (alternatives).

Double negation as in "I don't not like milk" rarely means literally "I do like milk" but rather conveys some sort of hedging, as though to imply that there is a third possibility. "Not not P" can be loosely interpreted as "surely P", and although P necessarily implies "not not P" the converse is suspect in English, much as with intuitionistic logic. In view of the highly idiosyncratic usage of conjunctions in natural languages, Boolean algebra cannot be considered a reliable framework for interpreting them.

Digital logic

Boolean operations are used in digital logic to combine the bits carried on individual wires, thereby interpreting them over {0,1}. When a vector of n identical binary gates are used to combine two bit vectors each of n bits, the individual bit operations can be understood collectively as a single operation on values from a Boolean algebra with 2^n elements.

Naive set theory

Naive set theory interprets Boolean operations as acting on subsets of a given set X. As we saw earlier this behavior exactly parallels the coordinate-wise combinations of bit vectors, with the union of two sets corresponding to the disjunction of two bit vectors and so on.

Video cards

The 256-element free Boolean algebra on three generators is deployed in computer displays based on raster graphics, which use bit blit to manipulate whole regions consisting of pixels, relying on Boolean operations to specify how the source region should be combined with the destination, typically with the help of a third region called the mask. Modern video cards offer all $2^{2^3} = 256$ ternary operations for this purpose, with the choice of operation being a one-byte (8-bit) parameter. The constants SRC = 0xaa or 10101010, DST = 0xcc or 11001100, and MSK = 0xf0 or 11110000 allow Boolean operations such as (SRC^DST)&MSK (meaning XOR the source and destination and then AND the result with the mask) to be written directly as a constant denoting a byte calculated at compile time, 0x80 in the (SRC^DST)&MSK example, 0x88 if just SRC^DST, etc. At run time the video card interprets the byte as the raster operation indicated by the original expression in a uniform way that requires remarkably little hardware and which takes time completely independent of the complexity of the expression.

Modeling and CAD

Solid modeling systems for computer aided design offer a variety of methods for building objects from other objects, combination by Boolean operations being one of them. In this method the space in which objects exist is understood as a set S of voxels (the three-dimensional analogue of pixels in two-dimensional graphics) and shapes are defined as subsets of S, allowing objects to be combined as sets via union, intersection, etc. One obvious use is in building a complex shape from simple shapes simply as the union of the latter.

Another use is in sculpting understood as removal of material: any grinding, milling, routing, or drilling operation that can be performed with physical machinery on physical materials can be simulated on the computer with the Boolean operation $x \wedge \neg y$ or x ″ y, which in set theory is set difference, remove the elements of y from those of x. Thus given two shapes one to be machined and the other the material to be removed, the result of machining the former to remove the latter is described simply as their set difference.

2

Boolean Algebra: Operations

There are only two values, and , unlike elementary algebra that deals with an infinity of values, the real numbers. Since there are only two values, a truth table is a very useful tool for working with Boolean algebra. A truth table lists all possible combinations of the variables in the problem. The resulting value of the Boolean operation(s) for each variable combination is shown on the respective row.

Boolean Algebra is a way of formally specifying, or describing, a particular situation or procedure. We use variables to represent elements of our situation or procedure. Variables may take one of only two values. Traditionally this would be True and False. So for instance we may have a variable X and state that this represents if it is raining outside or not. The value of X would be:

- True if it is raining outside.
- False if it is not raining outside.

What you have to remember is that although many things in the real world exist on a spectrum, in Boolean Algebra things are reduced to black and white. So we could have, for instance, light rain, steady rain, or heavy rain. In Boolean Algebra however, it is either raining or it isn't. This may seem a little limiting but this simplification of things actually turns out to be quite powerful.

It is possible to substitute other values in place of True and False. When working with computers it is often the case that True and False is replaced with 1 and 0. When working with physical circuits we may replace True and False with the presence or absence of a voltage.

In this way Boolean Algebra is useful to describe a process and then to build mechanisms which can perform those processes. Keep this in mind as you're working through the next few sections. This is what we are building towards.

BASIC OPERATIONS

We saw above that variables may be used to represent the current state of elements that we are interested in. Operations allow us to then define relationships between those variables. There are three basic operations. These are used often in Boolean expressions but are also used to create more complex operations. You'll probably find that you've actually used these operations quite a bit, you've just never thought of them formally before.

The result of an operation is similar to variables, it may only be either True or False.

It is chosen to always write operations in all UPPERCASE. This is so they are easily identified as operations. Many people follow this convention but it is not required. Feel free to use whatever method suits you best.

AND

The first operation is AND and it means pretty much what it does in plain english. So for instance I may state "If it's sunny outside AND I have completed my work then I will go for a run." To represent this in Boolean Algebra I may say that:

- x represents if it is sunny outside or not.
- y represents if I have completed my work or not.
- z represents if I go for a run or not.

And would write it like so:

x AND y = z

Here it is represented visually. The shaded region is the region which represents AND.

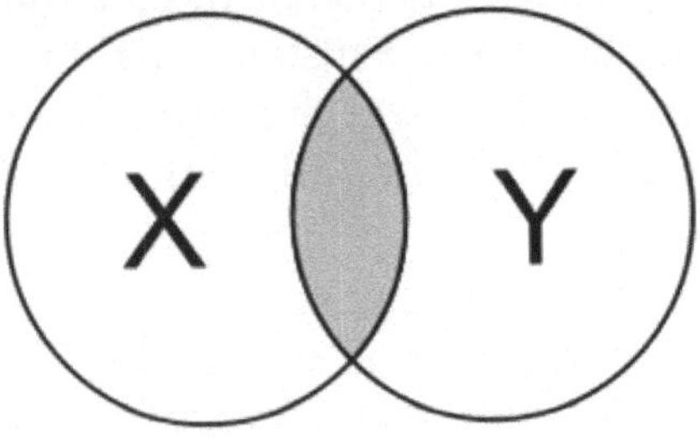

And now we'll represent it using what is called a Truth table. A truth table lists all the possible combinations of inputs for an expression (in this case a single operation) and what the result, or output should be.

X	Y	Result
False	False	False
True	False	False
False	True	False
True	True	True

OR

OR is also the same as how we would use it in plain english. It means that if either of the two variables is True then the result is True. So for instance I could say that "I will get home early from work if I get to leave early OR the traffic is goo≤.

Here is OR represented visually :

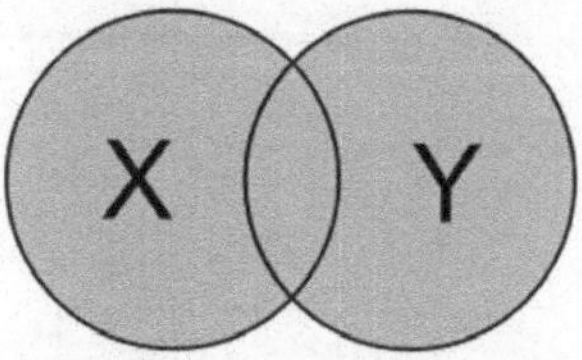

And again as a truth table :

X	Y	Result
False	False	False
True	False	True
False	True	True
True	True	True

Note that AND is False for all but True and True whilst OR is True for all but False and False. This observation will become useful to us later on. In fact there are many shortcuts and advantageous benefits to be gained from finding patterns like this so keep an eye out for them.

Not

Not is quite similar to how we use it in plain english. It has a subtle difference when used in Boolean Algebra. Normally I might say something like "I will eat dessert if I am not full". I could also have said "I will eat dessert if I am still hungry", which has the same meaning but using an opposite value. So not actually has the effect of flipping the value of a variable. If :

- the variable d currently has a value of True then

- the expression not d has a result of False
  Represented visually that is :

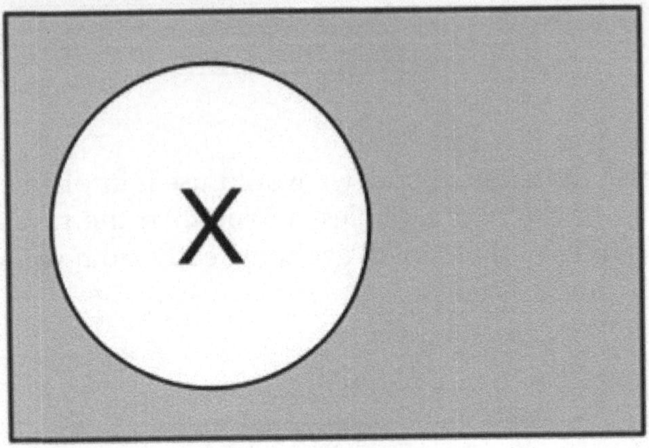

And as a truth table:

X	Result
True	False
False	True

Derived Operations

The above three operations are the building blocks for just about everything else we can do in Boolean Algebra. We will now introduce what are called derived operations. These are essentially shortcuts for commonly used combinations of the basic operations. As we will discover later on, some of these derived operations are very useful when we want to do computations and other things.

XOR or Exclusive OR

With the operation OR we saw that as long as one of the variables is True the result is True. It was also True if both of them were True. With the operation XOR we now say that the result will be True only if one of the two variables is True. That is, one of them is True but only one of them is True. We may build this operation from the basic operations like so :

g XOR p is equivalent to (g OR p) AND NOT(g AND p)

When brackets ( ) are used in an expression this means that we evaluate that part of the expression first before the other parts.

Let's run through an example to better understand what's going on.

If g is True and p is False then :

Substituting g and p for those values we get :

(True OR False) AND NOT(True AND False)

The first set of brackets (True OR False) AND NOT(True AND False) evaluates to True so let's replace that into the expression and we get :

True AND NOT(True AND False)

The next set of brackets True AND NOT(True AND False) evaluates to False so let's replace that into the expression as well giving us :

True AND NOT(False)

NOT(False) evaluates to True so we can apply that to the expression and we end up with :

True and True

And the final result is True.

Visually XOR looks like :

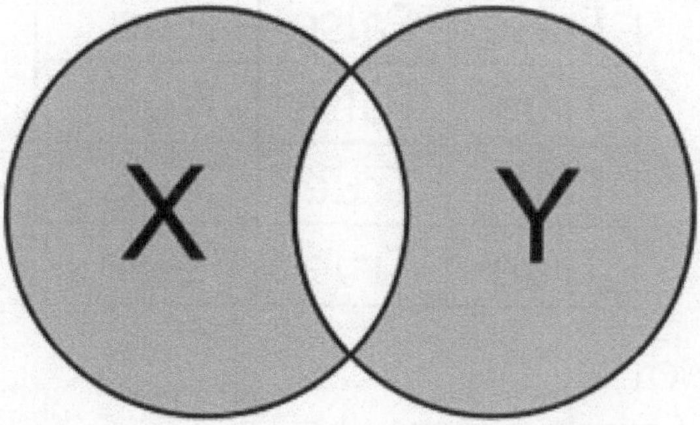

XOR as a truth table :

X	Y	Result
False	False	False
True	False	True
False	True	True
True	True	False

NAND or NOT AND

NAND is effectively the opposite of what AND is.

r NAND S is equivalent to NOT(r AND s)

Visually it looks like this :

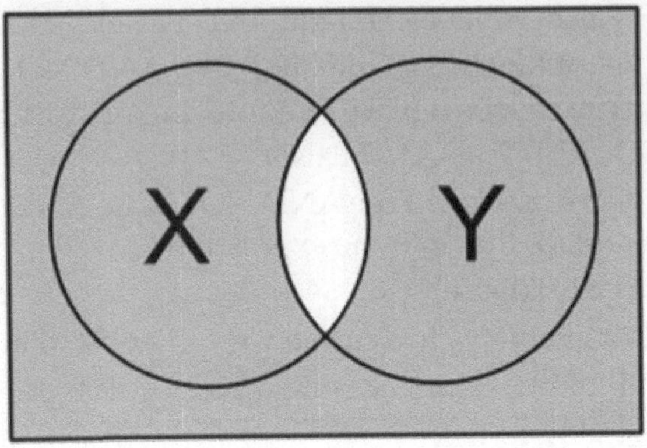

NAND as a truth table :

X	Y	Result
False	False	True
True	False	True
False	True	True
True	True	False

NOR or NOT OR
NOR is effectively the opposite of OR.
b NOR k is equivalent to NOT(b OR k)
Visually it looks like this :

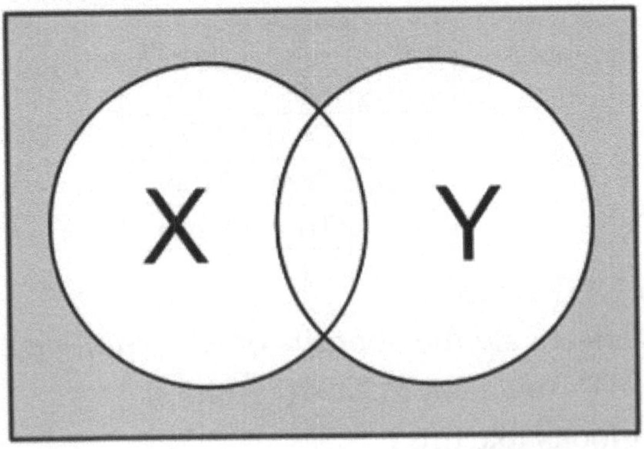

NOR as a truth table :

X	Y	Result
False	False	True
True	False	False
False	True	False
True	True	False

Boolean Algebra Postulates

Boolean algebra postulates are not laws or theorems but are statements that hold true. These postulates are the four possible logical OR and logical AND operations as well as the rules followed by the NOT operator. Given below are the boolean algebra postulates:

- $0 + 0 = 0$
- $0 + 1 = 1$
- $1 + 0 = 1$
- $1 + 1 = 1$
- $0 . 0 = 0$
- $0 . 1 = 0$
- $1 . 0 = 0$
- $1 . 1 = 1$
- $\overline{1} = 0$
- $\overline{0} = 1$

Axioms in Boolean Algebra

There are some set of logical expressions which we accept as true and upon which we can build a set of useful theorems. These sets of logical expressions are known as Axioms or postulates of Boolean Algebra. An axiom is nothing more than the definition of three basic logic operations (AND, OR and NOT). All axioms defined in boolean algebra are the results of an operation that is performed by a logical gate.

Axiom 1: 0.0 = 0 Axiom 6: 0+1 = 1

Axiom 2: 0.1 = 0 Axiom 7: 1+0 = 1

Axiom 3: 1.0 = 0 Axiom 8: 1+1 = 1

Axiom 4: 1.1 = 1 Axiom 9: 0 = 1

Axiom 5: 0+0 = 0 Axiom 10: 1 = 0

Based on these axioms we can conclude many laws of Boolean Algebra which are listed below,

Commutative Laws

A+B = B+A, and

A.B = B.A

Associative Laws

(A+B) + C = A+(B+C)

(A.B). C = A. (B.C)

AND Laws

A.0 = 0

A.1 = A

A.A = A

A.A = 0

OR Laws

A+0 = A

A+1 = 1

A+A = A

A+A = 0

Complementation Laws

If A = 0 then A = 1

If A = 1 then A = 0

A? = A

Distributive Laws

A(B+C) = AB + AC

A + BC = (A+B). (A+C)

Idempotence Law

A.A = A, If A=1, then A.A = 1.1 =1 = A and if A=0, then A.A = 0.0 = 0 = A

A+A = A, If A=1, then A+A = 1+1 =1 = A and if A=0, then A+A = 0+0 = 0 = A

Absorption Law

A + A.B = A

A.(A+B) = A

De-Morgan's Law

A+B = A. B

A.B = A + B

Consensus Theorem

A) AB + A C + BC = AB + A C

Proof:

LHS = AB + A C + BC
= AB + A C + BC (A+A)
= AB + A C + ABC + ABC
= AB (1+C) + A C (1+C)
= AB + A C = RHS

B) (A+B) (A + C) (B+C) = (A+B) (A + C)

Proof:

LHS = (A+B) (A + C) (B+C)
= (AA + AC + BA + BC) (B+C)
= (AC + BA + BC) (B+C)
= ABC + ACC + BAB + BAC + BCB + BCC
= ABC + AC + BA + BAC + BC + BC
= ABC + AC + BA + BAC + BC
= AC(B+1) + BA (1+C) + BC
= AC + BA + BC .... (Equation 1)

RHS = (A+B) (A + C)
= AA + AC + BA + BC
= AC + BA + BC .... (Equation 2)

Minimal axioms for Boolean algebra

In mathematical logic, minimal axioms for Boolean algebra are assumptions which are equivalent to the axioms of Boolean algebra (or propositional calculus), chosen to be as short as possible. For example, if one chooses to take commutativity for granted, an axiom with six NAND operations and three variables is equivalent to Boolean algebra :

((a #" b) #" c) #" (a #" ((a #" c) #" a)) = c

where the vertical bar represents the NAND logical operation (also known as the Sheffer stroke).

It is one of 25 candidate axioms for this property identified by Stephen Wolfram, by enumerating the Sheffer identities of length less or equal to 15 elements (excluding mirror images) that have no noncommutative models with four or fewer variables, and was first proven equivalent by William McCune, Branden Fitelson, and Larry Wos. MathWorld, a site associated with Wolfram, has named the axiom the "Wolfram axiom". McCune et al. also found a longer single axiom for Boolean algebra based on disjunction and negation.

In 1933, Edward Vermilye Huntington identified the axiom

¬ (¬ x (∨ y) (∨ ¬ (¬ x (∨ ¬ y) = x

as being equivalent to Boolean algebra, when combined with the commutativity of the OR operation, x (∨ y = y (∨ x, and the assumption of associativity, (x (∨ y) (∨ z = x (∨ (y (∨ z). Herbert Robbins conjectured that Huntington's axiom could be replaced by

¬ (¬ (x (∨ y) (∨ ¬ (x (∨ ¬ y)) = x,

which requires one fewer use of the logical negation operator . Neither Robbins nor Huntington could prove this conjecture; nor could Alfred Tarski, who took considerable interest in it later. The conjecture was eventually proved in 1996 with the aid of theorem-proving software. This proof established that the Robbins axiom, together with associativity and commutativity, form a 3-basis for Boolean algebra. The existence of a 2-basis was established in 1967 by Carew Arthur Meredith:

¬ (¬ x (∨ y) (∨ x = x,

¬ (¬ x (∨ ¬ y) (∨ (z (∨ y) = y (∨ (z (∨ x).

The following year, Meredith found a 2-basis in terms of the Sheffer stroke:

(x #" x) #" (y #" x) = x,

x | (y #" (x #" z)) = ((z #" y) #" y) #" x.

In 1973, Padmanabhan and Quackenbush demonstrated a method that, in principle, would yield a 1-basis for Boolean algebra. Applying this method in a straightforward manner yielded "axioms of enormous length", thereby prompting the question of how shorter axioms might be found. This search yielded the 1-basis in terms of the Sheffer stroke given above, as well as the 1-basis

¬ (¬ (¬ (x (" y) (" z) (" ¬ (x (" ¬ (¬ z (" ¬ (z (" u)))) = z,

which is written in terms of OR and NOT.

SUBALGEBRAS

Let A be a Boolean algebra and B a non-empty subset of A. Consider the following conditions:

1. 1. if a " B, then a2 " B,
2. if a,b " B, then a (" b " B,
3. if a,b " B, then a '" b " B,

It is easy to see that, by de Morgan's laws, conditions 1 and 2 imply conditions 1 and 3, and vice versa. Also, If B satisfies 1 and 2, then 1 = a (" a2 " B by picking an element a " B. As a result, 0 = 12 " B as well.

A non-empty subset B of a Boolean algebra A satisfying conditions 1 and 2 (or equivalently 1 and 3) is called a Boolean subalgebra of A.

Examples.

- Every Boolean algebra contains a unique two-element Boolean algebra consisting of just 0 and 1.
- If A contains a non-trivial element a (‘“0), then 0,1,a,a2 form a four-element Boolean subalgebra of A.
- Here is a concrete example. Let A be the field of all sets of a set X (the power set of X). Let B be the set of all finite and all cofinite subsets of A. Then B is a subalgebra of A.
- Continuing from the example above, if X is equipped with a topology T, then the set of all clopen sets in X is a Boolean subalgebra of the algebra of all regular open sets of X.

Remarks

Let A be a Boolean algebra.

- Arbitrary intersections of Boolean subalgebras of A is a Boolean subalgebra.
- An arbitrary union of an increasing chain of Boolean subalgebras of A is a Boolean subalgebra.
- A subalgebra B of A is said to be dense in A if it is dense as a subset of the underlying poset A. In other words, B is dense in A iff for any a″A, there is a b″B such that b≤a. It can be easily shown that B is dense in A is equivalent to any one of the following:
  a. for any a ″ A, there is b″B such that a ≤ b;
  b. for any x,y ″ A with x ≤ y, there is z ″ B such that x≤z≤y.

To see the last equivalence, notice first that by picking x = 0 we see that B is dense, and conversely, if x ≤ y, then there is r ″ B such that r≤y, so that x ≤ x (″ r ≤ y.

Subalgebras Generated by a Set

Let A be a Boolean algebra and X a subset of A. The intersection of all Boolean subalgebras (of A) containing X is called the Boolean subalgebra generated by X, and is denoted by è′Xé′. As indicated by the remark above, è′Xé′ is a Boolean subalgebra of A, the smallest subalgebra containing X. If è′Xé′=A, then we say that the Boolean algebra A itself is generated by X.

Proposition 1. Every element of è′Xé′⊆A has a disjunctive normal form (DNF). In other words, if a″è′Xé′, then

$$a = \underset{n}{\overset{i=1}{\square}} \underset{\phi(i)}{\overset{j=1}{\square}} a_{ij} = (a_{11}\wedge\cdots\wedge a_{1\phi(1)})\vee(a_{21}\wedge\cdots\wedge a_{2\phi(2)})\vee\cdots\vee(a_{n1}\wedge\cdots\wedge a_{n\phi(n)}),$$

where either a_{ij} or $a_{2\ ij}$ belongs to X.

Boolean Expressions and Functions

In mathematics, a Boolean function is a function whose arguments and result assume values from a two-element set (usually {true, false}, {0,1} or {-1,1}). Alternative names are switching function, used especially in older computer science literature, and truth function (or logical function), used in logic. Boolean functions are the subject of Boolean algebra and switching theory.

A Boolean function takes the form f : { 0 , 1 } k → { 0 , 1 }, where { 0 , 1 } is known as the Boolean domain and k is a non-negative integer called the arity of the function. In the case where k = 0, the function is a constant element of { 0, 1}. A Boolean function with multiple outputs, f : { 0 , 1 } k → { 0 , 1 } m with m > 1 is a vectorial or vector-valued Boolean function (an S-box in cryptography).

There are 2^{2k} different Boolean functions with k arguments; equal to the number of different truth tables with 2 k entries.

Every k-ary Boolean function can be expressed as a propositional formula in k variables $x_1, \ldots, x_k$, and two propositional formulas are logically equivalent if and only if they express the same Boolean function.

Boolean algebra is algebra of logic. It deals with variables that can have two discrete values, 0 (False) and 1 (True); and operations that have logical significance. The earliest method of manipulating symbolic logic was invented by George Boole and subsequently came to be known as Boolean Algebra.

Boolean algebra has now become an indispensable tool in computer science for its wide applicability in switching theory, building basic electronic circuits and design of digital computers.

Examples

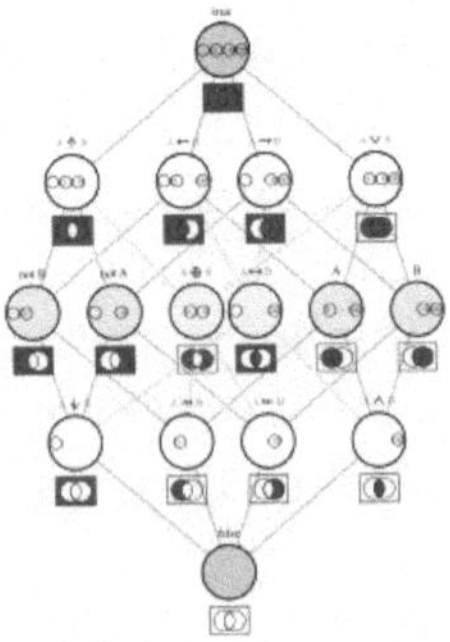

The sixteen binary Boolean functions

The rudimentary symmetric Boolean functions (logical connectives or logic gates) are:

- NOT, negation or complement - which receives one input and returns true when that input is false ("not")
- AND or conjunction - true when all inputs are true ("both")
- OR or disjunction - true when any input is true ("either")
- XOR or exclusive disjunction - true when one of its inputs is true and the other is false ("not equal")
- NAND or Sheffer stroke - true when it is not the case that all inputs are true ("not both")
- NOR or logical nor - true when none of the inputs are true ("neither")
- XNOR or logical equality - true when both inputs are the same ("equal")

An example of a more complicated function is the majority function (of an odd number of inputs).

Representation

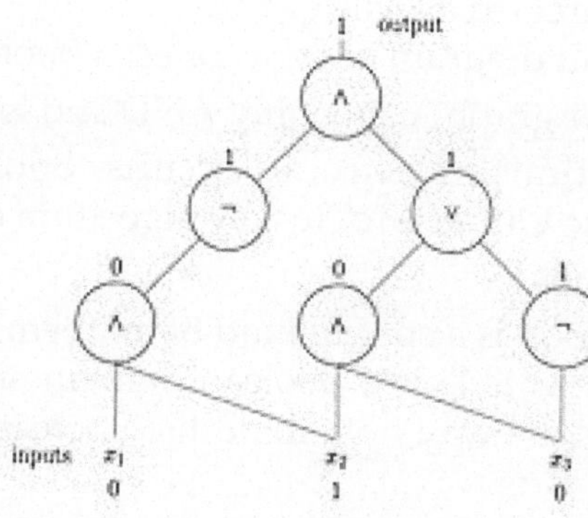

A Boolean function represented as a Boolean circuit

A Boolean function may be specified in a variety of ways:

- Truth table: explicitly listing its value for all possible values of the arguments
  - o Marquand diagram: truth table values arranged in a two-dimensional grid (used in a Karnaugh map)
  - o Binary decision diagram, listing the truth table values at the bottom of a binary tree
  - o Venn diagram, depicting the truth table values as a colouring of regions of the plane

Algebraically, as a propositional formula using rudimentary boolean functions:

- Negation normal form, an arbitrary mix of AND and ORs of the arguments and their complements
- Disjunctive normal form, as an OR of ANDs of the arguments and their complements
- Conjunctive normal form, as an AND of ORs of the arguments and their complements
- Canonical normal form, a standardized formula which uniquely identifies the function:

o Algebraic normal form or Zhegalkin polynomial, as a XOR of ANDs of the arguments (no complements allowed)

o Full (canonical) disjunctive normal form, an OR of ANDs each containing every argument or complement (minterms)

o Full (canonical) conjunctive normal form, an AND of ORs each containing every argument or complement (maxterms)

o Blake canonical form, the OR of all the prime implicants of the function

Boolean formulas can also be displayed as a graph:

- Propositional directed acyclic graph
  - o Digital circuit diagram of logic gates, a Boolean circuit
  - o And-inverter graph, using only AND and NOT

In order to optimize electronic circuits, Boolean formulas can be minimized using the Quine–McCluskey algorithm or Karnaugh map.

Boolean Functions

A Boolean function is a special kind of mathematical function f:Xn→X of degree n, where X={0,1} is a Boolean domain and n is a non-negative integer. It describes the way how to derive Boolean output from Boolean inputs.

Example " Let, F(A,B)=A2 B2 . This is a function of degree 2 from the set of ordered pairs of Boolean variables to the set {0,1} where F(0,0)=1,F(0,1)=0,F(1,0)=0 and F(1,1)=0

Representation of Boolean Functions

A Boolean function is described by an algebraic expression consisting of binary variables, the constants 0 and 1, and the logic operation symbols $+, ., '$. For a given set of values of the binary variables involved, the boolean function can have a value of 0 or 1. For example, the boolean function $F = x'y + z$ is defined in terms of three binary variables x, y, z. The function is equal to 1 if $x = 0$ and $y = 1$ simultaneously or $z = 1$.

Every boolean function can be expressed by an algebraic expression, such as one mentioned above, or in terms of a Truth Table. A function may be expressed through several algebraic expressions, on account of them being logically equivalent, but there is only one unique truth table for every function.

A Boolean function can be transformed from an algebraic expression into a circuit diagram composed of logic gates connected in a particular structure. Circuit diagram for F⁻

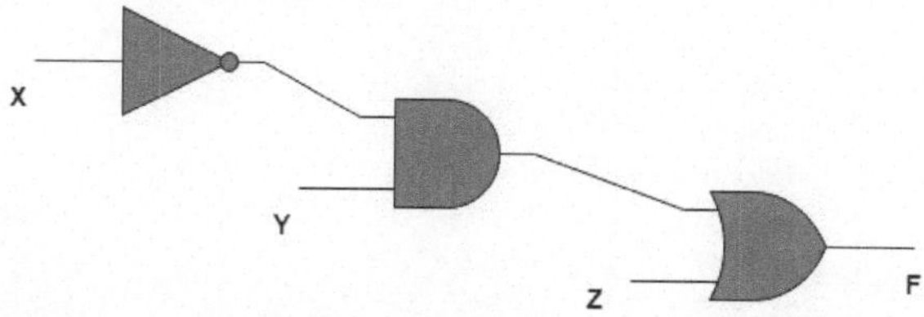

Canonical and Standard Forms

Any binary variable can take one of two forms, x or x′. A boolean function can be expressed in terms of n binary variables. If all the binary variables are combined together using the AND operation, then there are a total of 2^n combinations since each variable can take two forms.

Each of the combinations is called a minterm or standard product. A minterm is represented by m_i where i is the decimal equivalent of the binary number the minterm is designated.

Important Note – In a minterm, the binary variable is un-primed if the variable is 1 and it is primed if the variable is 0 i.e. if the minterm is xy′ then that means x = 1 and y = 0.

For example, for a boolean function in two variables the minterms are –

$$m_0 = x'y', m_1 = x'y, m_2 = xy', m_3 = xy$$

In a similar way, if the variables are combined together with OR operation, then the term obtained is called a maxterm or standard sum. A maxterm is represented by M_i where i is the decimal equivalent of the binary number the maxterm is designated.

Boolean Expressions

A Boolean expression always produces a Boolean value. A Boolean expression is composed of a combination of the Boolean constants (True or False), Boolean variables and logical connectives. Each Boolean expression represents a Boolean function.

Example " AB2 C is a Boolean expression.

Boolean Identities

Double Complement Law

<“(<“A)=A

Complement Law

A+<“A=1

(OR Form)

A.<“A=0

(AND Form)

Idempotent Law

A+A=A

(OR Form)

A.A=A

(AND Form)

Identity Law

A+0=A

(OR Form)

A.1=A

(AND Form)

Dominance Law

A+1=1

(OR Form)

A.0=0

(AND Form)

Commutative Law

A+B=B+A

(OR Form)

A.B=B.A

(AND Form)

Associative Law

A+(B+C)=(A+B)+C

(OR Form)

A.(B.C)=(A.B).C

(AND Form)

Absorption Law

A.(A+B)=A

A+(A.B)=A

Simplification Law

A.(<“A+B)=A.B

A+(<“A.B)=A+B

Distributive Law

A+(B.C)=(A+B).(A+C)

A.(B+C)=(A.B)+(A.C)

De-Morgan's Law

<“(A.B)=<“A+<“B

<“(A+B)=<“A.<“B

ANALYSIS

Properties

A Boolean function can have a variety of properties:

- Constant: Is always true or always false regardless of its arguments.
- Monotone: for every combination of argument values, changing an argument from false to true can only cause the output to switch from false to true and not from true to false. A function is said to be unate in a certain variable if it is monotone with respect to changes in that variable.
- Linear: for each variable, flipping the value of the variable either always makes a difference in the truth value or never makes a difference (a parity function).
- Symmetric: the value does not depend on the order of its arguments.
- Read-once: Can be expressed with conjunction, disjunction, and negation with a single instance of each variable.
- Balanced: if its truth table contains an equal amount of zeros and ones. The Hamming weight of the function is the number of ones in the truth table.
- Bent: its derivatives are all balanced (the autocorrelation spectrum is zero)
- Correlation immune to mth order: if the output is uncorrelated with all (linear) combinations of at most m arguments
- Evasive: if evaluation of the function always requires the value of all arguments
- A Boolean function is a Sheffer function if it can be used to create (by composition) any arbitrary Boolean function (see functional completeness)
- The algebraic degree of a function is the order of the highest order monomial in its algebraic normal form

Circuit complexity attempts to classify Boolean functions with respect to the size or depth of circuits that can compute them.

DERIVED FUNCTIONS

A Boolean function may be decomposed using Boole's expansion theorem in positive and negative Shannon cofactors (Shannon expansion), which are the (k-1)-ary functions resulting from fixing one of the arguments (to zero or one). The general (k-ary) functions obtained by imposing a linear constraint on a set of inputs (a linear subspace) are known as subfunctions.

The Boolean derivative of the function to one of the arguments is a (k-1)-ary function that is true when the output of the function is sensitive to the chosen input variable; it is the XOR of the two corresponding cofactors. A derivative and a cofactor are used in a Reed–Muller expansion. The concept can be generalized as a k-ary derivative in the direction dx, obtained as the difference (XOR) of the function at x and x + dx.

The Möbius transform (or Boole-Möbius transform) of a Boolean function is the set of coefficients of its polynomial (algebraic normal form), as a function of the monomial exponent vectors. It is a self-inverse transform. It can be calculated efficiently using a butterfly algorithm ("Fast Möbius Transform"), analogous to the Fast Fourier Transform. Coincident Boolean functions are equal to their Möbius transform, i.e. their truth table (minterm) values equal their algebraic (monomial) coefficients. There are 2^2^(k"1) coincident functions of k arguments.

Cryptographic analysis

The Walsh transform of a Boolean function is a k-ary integer-valued function giving the coefficients of a decomposition into linear functions (Walsh functions), analogous to the decomposition of real-valued functions into harmonics by the Fourier transform. Its square is the power spectrum or Walsh spectrum. The Walsh coefficient of a single bit vector is a measure for the correlation of that bit with the output of the Boolean function. The maximum (in absolute value) Walsh coefficient is known as the linearity of the function. The highest number of bits (order) for which all Walsh coefficients are 0 (i.e. the subfunctions are balanced) is known as resiliency, and the function is said to be correlation immune to that order. The Walsh coefficients play a key role in linear cryptanalysis.

The autocorrelation of a Boolean function is a k-ary integer-valued function giving the correlation between a certain set of changes in the inputs and the function ouput. For a given bit vector it is related to the Hamming weight of the derivative in that direction. The maximal autocorrelation coefficient (in absolute value) is known as the absolute indicator. If all autocorrelation coefficients are 0 (i.e. the derivatives are balanced) for a certain number of bits then the function is said to satisfy the propagation criterion to that order; if they are all zero then the function is a bent function. The autocorrelation coefficients play a key role in differential cryptanalysis.

The Walsh coefficients of a Boolean function and its autocorrelation coefficients are related by the equivalent of the Wiener–Khinchin theorem, which states that the autocorrelation and the power spectrum are a Walsh transform pair.

These concepts can be extended naturally to vectorial Boolean functions by considering their output bits (coordinates) individually, or more thoroughly, by looking at the set of all linear functions of output bits, known as its components. The set of Walsh transforms of the components is known as a Linear Approximation Table (LAT) or correlation matrix; it describes the correlation between different linear combinations of input and output bits. The set of autocorrelation coefficients of the components is the autocorrelation table, related by a Walsh transform of the components to the more widely used Difference Distribution Table (DDT) which lists the correlations between differences in input and output bits.

Applications

Boolean functions play a basic role in questions of complexity theory as well as the design of processors for digital computers, where they are implemented in electronic circuits using logic gates.

The properties of Boolean functions are critical in cryptography, particularly in the design of symmetric key algorithms (see substitution box).

In cooperative game theory, monotone Boolean functions are called simple games (voting games); this notion is applied to solve problems in social choice theory.

Truth Table Formation

A truth table represents a table having all combinations of inputs and their corresponding result.

It is possible to convert the switching equation into a truth table. For example, consider the following switching equation.

$$F(A, B, C) = A + BC$$

The output will be high (1) if A = 1 or BC = 1 or both are 1. The truth table for this equation is shown by Table (a). The number of rows in the truth table is 2^n where n is the number of input variables (n=3 for the given equation). Hence there are $2^3 = 8$ possible input combination of inputs.

Inputs			Output
A	B	C	F
0	0	0	0
0	0	1	0
0	1	0	0
0	1	1	1
1	0	0	1
1	0	1	1
1	1	0	1
1	1	1	1

Methods to simplify the boolean function

The methods used for simplifying the Boolean function are as follows "

- Karnaugh-map or K-map, and
- NAND gate method.

Karnaugh-map or K-map

The Boolean theorems and the De-Morgan's theorems are useful in manipulating the logic expression. We can realize the logical expression using gates. The number of logic gates required for the realization of a logical expression should be reduced to a minimum possible value by K-map method. This method can be done in two different ways, as discussed below.

Sum of Products (SOP) Form

It is in the form of sum of three terms AB, AC, BC with each individual term is a product of two variables. Say A.B or A.C etc. Therefore such expressions are known as expression in SOP form. The sum and products in SOP form are not the actual additions or multiplications. In fact they are the OR and AND functions. In SOP form, 0 represents a bar and 1 represents an unbar. SOP form is represented by Σ.

Given below is an example of SOP.

In SOP form $\overline{A}\,\overline{B} + \overline{A}B + AB$

$\overline{A}\,\overline{B} \rightarrow 00$, $\overline{A}B \rightarrow 01$, $AB \rightarrow 11$

A \ B	0	1
0	1	1
1		1

$\sum m(0,1,3)$

Answer: $\overline{A}\,\overline{B} + \overline{A}B + AB = \overline{A} + B$

Product of Sums (POS) Form

It is in the form of product of three terms (A+B), (B+C), or (A+C) with each term is in the form of a sum of two variables. Such expressions are said to be in the product of sums (POS) form. In POS form, 0 represents an unbar and 1 represents a bar.

Given below is an example of POS.

In POS form $(B + \overline{C})(\overline{A} + \overline{B})(\overline{B} + C)$

$\downarrow$ 0 1 1 1 1 0

A \ BC	00	01	10	11
0		0	0	0
1				

$\prod m(1,3,2)$

Answer : $(A + \overline{C})(A + \overline{B})$

NAND gates Realization

NAND gates can be used to simplify Boolean functions as shown in the example below.

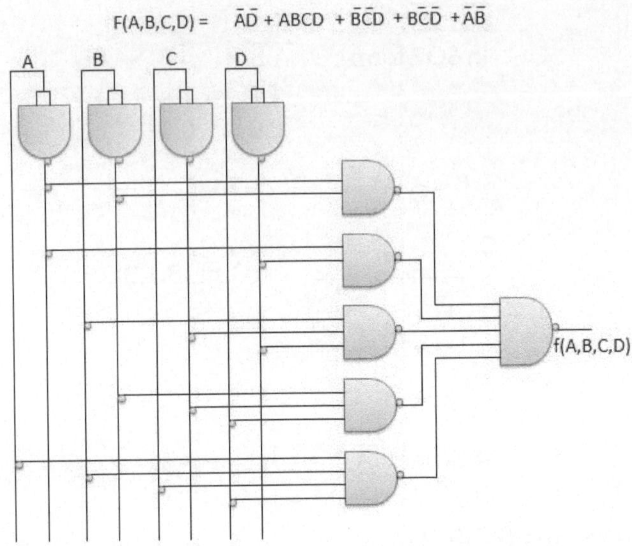

Normal Forms

In Boolean algebra, the algebraic normal form (ANF), ring sum normal form (RSNF or RNF), Zhegalkin normal form, or Reed–Muller expansion is a way of writing logical formulas in one of three subforms:

- The entire formula is purely true or false:
  1
  0
- One or more variables are ANDed together into a term, then one or more terms are XORed together into ANF. No NOTs are permitted:
  $a \oplus b \oplus ab \oplus abc$
  or in standard propositional logic symbols:
  a »″ b »″ (a ‴ b) »″ (a ‴ b ‴ c)
- The previous subform with a purely true term:
  $1 \oplus a \oplus b \oplus ab \oplus abc$

Formulas written in ANF are also known as Zhegalkin polynomials and Positive Polarity (or Parity) Reed–Muller expressions (PPRM).

Common uses

ANF is a normal form, which means that two equivalent formulas will convert to the same ANF, easily showing whether two formulas are equivalent for automated theorem proving. Unlike other normal forms, it can be represented as a simple list of lists of variable names; conjunctive and disjunctive normal forms also require recording whether each variable is negated or not. Negation normal form is unsuitable for that purpose, since it doesn't use equality as its equivalence relation: a (" ¬a isn't reduced to the same thing as 1, even though they're equal.

Putting a formula into ANF also makes it easy to identify linear functions (used, for example, in linear-feedback shift registers): a linear function is one that is a sum of single literals. Properties of nonlinear-feedback shift registers can also be deduced from certain properties of the feedback function in ANF.

Performing operations within algebraic normal form

There are straightforward ways to perform the standard boolean operations on ANF inputs in order to get ANF results.

XOR (logical exclusive disjunction) is performed directly:

$(1 \oplus x) \oplus (1 \oplus x \oplus y)$

$1 \oplus x \oplus 1 \oplus x \oplus y$

$1 \oplus 1 \oplus x \oplus x \oplus y$

y

NOT (logical negation) is XORing 1:

$\neg(1 \oplus x \oplus y)$

$1 \oplus (1 \oplus x \oplus y)$

$1 \oplus 1 \oplus x \oplus y$

$x \oplus y$

AND (logical conjunction) is distributed algebraically

$(1 \oplus x)(1 \oplus x \oplus y)$

$1(1 \oplus x \oplus y) \oplus x(1 \oplus x \oplus y)$

$(1 \oplus x \oplus y) \oplus (x \oplus x \oplus xy)$

$1 \oplus x \oplus x \oplus x \oplus y \oplus xy$

$1 \oplus x \oplus y \oplus xy$

OR (logical disjunction) uses either $1 \oplus (1 \oplus a)(1 \oplus b)$ (easier when both operands have purely true terms) or $a \oplus b \oplus ab$ (easier otherwise):

$(1 \oplus x) + (1 \oplus x \oplus y)$

$1 \oplus (1 \oplus 1 \oplus x)(1 \oplus 1 \oplus x \oplus y)$

$1 \oplus x(x \oplus y)$

$1 \oplus x \oplus xy$

Converting to algebraic normal form

Each variable in a formula is already in pure ANF, so you only need to perform the formula's boolean operations as shown above to get the entire formula into ANF. For example:

$x + (y . \neg z)$

$x + (y(1 \oplus z))$

$x + (y \oplus yz)$

$x \oplus (y \oplus yz) \oplus x(y \oplus yz)$

$x \oplus y \oplus xy \oplus yz \oplus xyz$

BOOLEAN ALGEBRAS AND PROPOSITIONAL LOGIC

Propositional calculus is a branch of logic. It is also called propositional logic, statement logic, sentential calculus, sentential logic, or sometimes zeroth-order logic. It deals with propositions (which can be true or false) and relations between propositions, including the construction of arguments based on them. Compound propositions are formed by connecting propositions by logical connectives. Propositions that contain no logical connectives are called atomic propositions.

Unlike first-order logic, propositional logic does not deal with non-logical objects, predicates about them, or quantifiers. However, all the machinery of propositional logic is included in first-order logic and higher-order logics. In this sense, propositional logic is the foundation of first-order logic and higher-order logic.

A Proposition is an atomic sentence that can either be TRUE or FALSE(and nothing else.

Following sentences are example of proposition.

Proposition: India is a country

Result: TRUE

Proposition: 100 is greater than 200

Result: FALSE

Whereas the sentence How are you? is not a proposition as the answer can't be TRUE or FALSE.

Explanation

Logical connectives are found in natural languages. In English for example, some examples are "an≤ (conjunction), "or" (disjunction), "not" (negation) and "if" (but only when used to denote material conditional).

The following is an example of a very simple inference within the scope of propositional logic:

Premise 1: If it's raining then it's cloudy.

Premise 2: It's raining.

Conclusion: It's cloudy.

Both premises and the conclusion are propositions. The premises are taken for granted, and with the application of modus ponens (an inference rule), the conclusion follows. As propositional logic is not concerned with the structure of propositions beyond the point where they can't be decomposed any more by logical connectives, this inference can be restated replacing those atomic statements with statement letters, which are interpreted as variables representing statements:

Premise 1: $P \rightarrow Q$

Premise 2: P

Conclusion: Q

The same can be stated succinctly in the following way:

$P \rightarrow Q$, P ¢″ Q

When P is interpreted as "It's raining" and Q as "it's cloudy" the above symbolic expressions can be seen to correspond exactly with the original expression in natural language. Not only that, but they will also correspond with any other inference of this form, which will be valid on the same basis this inference is.

Propositional logic may be studied through a formal system in which formulas of a formal language may be interpreted to represent propositions. A system of axioms and inference rules allows certain formulas to be derived. These derived formulas are called theorems and may be interpreted to be true propositions. A constructed sequence of such formulas is known as a derivation or proof and the last formula of the sequence is the theorem. The derivation may be interpreted as proof of the proposition represented by the theorem.

When a formal system is used to represent formal logic, only statement letters (usually capital roman letters such as P, Q and R) are represented directly. The natural language propositions that arise when they're interpreted are outside the scope of the system, and the relation between the formal system and its interpretation is likewise outside the formal system itself.

In classical truth-functional propositional logic, formulas are interpreted as having precisely one of two possible truth values, the truth value of true or the truth value of false. The principle of bivalence and the law of excluded middle are upheld. Truth-functional propositional logic defined as such and systems isomorphic to it are considered to be zeroth-order logic. However, alternative propositional logics are also possible. For more, see Other logical calculi below.

Although propositional logic (which is interchangeable with propositional calculus) had been hinted by earlier philosophers, it was developed into a formal logic (Stoic logic) by Chrysippus in the 3rd century BC and expanded by his successor Stoics. The logic was focused on propositions. This advancement was different from the traditional syllogistic logic, which was focused on terms. However, most of the original writings were lost and the propositional logic developed by the Stoics was no longer understood later in antiquity.[citation needed] Consequently, the system was essentially reinvented by Peter Abelard in the 12th century.

Propositional logic was eventually refined using symbolic logic. The 17th/18th-century mathematician Gottfried Leibniz has been credited with being the founder of symbolic logic for his work with the calculus ratiocinator. Although his work was the first of its kind, it was unknown to the larger logical community. Consequently, many of the advances achieved by Leibniz were recreated by logicians like George Boole and Augustus De Morgan—completely independent of Leibniz.

Just as propositional logic can be considered an advancement from the earlier syllogistic logic, Gottlob Frege's predicate logic can be also considered an advancement from the earlier propositional logic. One author describes predicate logic as combining "the distinctive features of syllogistic logic and propositional logic." Consequently, predicate logic ushered in a new era in logic's history; however, advances in propositional logic were still made after Frege, including natural deduction, truth trees and truth tables. Natural deduction was invented by Gerhard Gentzen and Jan £ukasiewicz. Truth trees were invented by Evert Willem Beth. The invention of truth tables, however, is of uncertain attribution.

Within works by Frege and Bertrand Russell, are ideas influential to the invention of truth tables. The actual tabular structure (being formatted as a table), itself, is generally credited to either Ludwig Wittgenstein or Emil Post (or both, independently). Besides Frege and Russell, others credited with having ideas preceding truth tables include Philo, Boole, Charles Sanders Peirce, and Ernst Schröder. Others credited with the tabular structure include Jan £ukasiewicz, Alfred North Whitehead, William Stanley Jevons, John Venn, and Clarence Irving Lewis. Ultimately, some have concluded, like John Shosky, that "It is far from clear that any one person should be given the title of 'inventor' of truth-tables.".

In general terms, a calculus is a formal system that consists of a set of syntactic expressions (well-formed formulas), a distinguished subset of these expressions (axioms), plus a set of formal rules that define a specific binary relation, intended to be interpreted as logical equivalence, on the space of expressions.

When the formal system is intended to be a logical system, the expressions are meant to be interpreted as statements, and the rules, known to be inference rules, are typically intended to be truth-preserving. In this setting, the rules, which may include axioms, can then be used to derive ("infer") formulas representing true statements—from given formulas representing true statements.

The set of axioms may be empty, a nonempty finite set, or a countably infinite set (see axiom schema). A formal grammar recursively defines the expressions and well-formed formulas of the language. In addition a semantics may be given which defines truth and valuations (or interpretations).

The language of a propositional calculus consists of

1. a set of primitive symbols, variously referred to as atomic formulas, placeholders, proposition letters, or variables, and
2. a set of operator symbols, variously interpreted as logical operators or logical connectives.

A well-formed formula is any atomic formula, or any formula that can be built up from atomic formulas by means of operator symbols according to the rules of the grammar.

Mathematicians sometimes distinguish between propositional constants, propositional variables, and schemata. Propositional constants represent some particular proposition, while propositional variables range over the set of all atomic propositions. Schemata, however, range over all propositions. It is common to represent propositional constants by A, B, and C, propositional variables by P, Q, and R, and schematic letters are often Greek letters, most often ö, ø, and ÷.

Basic concepts

The following outlines a standard propositional calculus. Many different formulations exist which are all more or less equivalent, but differ in the details of:

1. their language (i.e., the particular collection of primitive symbols and operator symbols),
2. the set of axioms, or distinguished formulas, and
3. the set of inference rules.

Any given proposition may be represented with a letter called a 'propositional constant', analogous to representing a number by a letter in mathematics (e.g., a = 5). All propositions require exactly one of two truth-values: true or false. For example, let P be the proposition that it is raining outside. This will be true (P) if it is raining outside, and false otherwise (¬P).

- We then define truth-functional operators, beginning with negation. ¬P represents the negation of P, which can be thought of as the denial of P. In the example above, ¬P expresses that it is not raining outside, or by a more standard reading: "It is not the case that it is raining outside." When P is true, ¬P is false; and when P is false, ¬P is true. As a result, ¬ ¬P always has the same truth-value as P.

- Conjunction is a truth-functional connective which forms a proposition out of two simpler propositions, for example, P and Q. The conjunction of P and Q is written P 'H Q, and expresses that each are true. We read P 'H Q as "P and Q". For any two propositions, there are four possible assignments of truth values:

1. P is true and Q is true
2. P is true and Q is false
3. P is false and Q is true
4. P is false and Q is false

The conjunction of P and Q is true in case 1, and is false otherwise. Where P is the proposition that it is raining outside and Q is the proposition that a cold-front is over Kansas, P 'H Q is true when it is raining outside and there is a cold-front over Kansas. If it is not raining outside, then P 'H Q is false; and if there is no cold-front over Kansas, then P 'H Q is also false.

- Disjunction resembles conjunction in that it forms a proposition out of two simpler propositions. We write it P (H Q, and it is read "P or Q". It expresses that either P or Q is true. Thus, in the cases listed above, the disjunction of P with Q is true in all cases—except case 4. Using the example above, the disjunction expresses that it is either raining outside, or there is a cold front over Kansas. (Note, this use of disjunction is supposed to resemble the use of the English word "or". However, it is most like the English inclusive "or", which can be used to express the truth of at least one of two propositions. It is not like the English exclusive "or", which expresses the truth of exactly one of two propositions. In other words, the exclusive "or" is false when both P and Q are true (case 1). An example of the exclusive or is: You may have a bagel or a pastry, but not both. Often in natural language, given the appropriate context, the addendum "but not both" is omitted—but implied. In mathematics, however, "or" is always inclusive or; if exclusive or is meant it will be specified, possibly by "xor".)
- Material conditional also joins two simpler propositions, and we write $P \rightarrow Q$, which is read "if P then Q". The proposition to the left of the arrow is called the antecedent, and the proposition to the right is called the consequent. (There is no such designation for conjunction or disjunction, since they are commutative operations.) It expresses that Q is true whenever P is true. Thus $P \rightarrow Q$ is true in every case above except case 2, because this is the only case when P is true but Q is not. Using the example, if P then Q expresses that if it is raining outside, then there is a cold-front over Kansas. The material conditional is often confused with physical causation. The material conditional, however,

only relates two propositions by their truth-values—which is not the relation of cause and effect. It is contentious in the literature whether the material implication represents logical causation.

- Biconditional joins two simpler propositions, and we write P "! Q, which is read "P if and only if Q". It expresses that P and Q have the same truth-value, and in cases 1 and 4. 'P is true if and only if Q' is true, and is false otherwise.

It is very helpful to look at the truth tables for these different operators, as well as the method of analytic tableaux.

CLOSURE UNDER OPERATIONS

Propositional logic is closed under truth-functional connectives. That is to say, for any proposition ö, ¬ö is also a proposition. Likewise, for any propositions ö and ø, ö '" ø is a proposition, and similarly for disjunction, conditional, and biconditional. This implies that, for instance, ö '" ø is a proposition, and so it can be conjoined with another proposition. In order to represent this, we need to use parentheses to indicate which proposition is conjoined with which. For instance, P '" Q '" R is not a well-formed formula, because we do not know if we are conjoining P '" Q with R or if we are conjoining P with Q '" R.

Thus, we must write either (P '" Q) '" R to represent the former, or P '" (Q '" R) to represent the latter. By evaluating the truth conditions, we see that both expressions have the same truth conditions (will be true in the same cases), and moreover that any proposition formed by arbitrary conjunctions will have the same truth conditions, regardless of the location of the parentheses. This means that conjunction is associative, however, one should not assume that parentheses never serve a purpose. For instance, the sentence P '" (Q (" R) does not have the same truth conditions of (P '" Q) (" R, so they are different sentences distinguished only by the parentheses. One can verify this by the truth-table method referenced above.

Note: For any arbitrary number of propositional constants, we can form a finite number of cases which list their possible truth-values. A simple way to generate this is by truth-tables, in which one writes P, Q, ..., Z, for any list of k propositional constants—that is to say, any list of propositional constants with k entries. Below this list, one writes 2^k rows, and below P one fills in the first half of the rows with true (or T) and the second half with false (or F). Below Q one fills in one-quarter of the rows with T, then one-quarter with F, then one-quarter with T and the last quarter with F. The next column alternates between true and false for each eighth of the rows, then sixteenths, and so on, until the last propositional constant varies between T and F for each row. This will give a complete listing of cases or truth-value assignments possible for those propositional constants.

Soundness and completeness of the rules

The crucial properties of this set of rules are that they are sound and complete. Informally this means that the rules are correct and that no other rules are required. These claims can be made more formal as follows. Note that the proofs for the soundness and completeness of the propositional logic are not themselves proofs in propositional logic ; these are theorems in ZFC used as a metatheory to prove properties of propositional logic.

We define a truth assignment as a function that maps propositional variables to true or false. Informally such a truth assignment can be understood as the description of a possible state of affairs (or possible world) where certain statements are true and others are not. The semantics of formulas can then be formalized by defining for which "state of affairs" they are considered to be true, which is what is done by the following definition.

We define when such a truth assignment A satisfies a certain well-formed formula with the following rules:

- A satisfies the propositional variable P if and only if A(P) = true
- A satisfies ¬ö if and only if A does not satisfy ö
- A satisfies (φ ∧ ψ) if and only if A satisfies both ö and ø
- A satisfies (φ (∨ ψ) if and only if A satisfies at least one of either ö or ø
- A satisfies (φ → ψ) if and only if it is not the case that A satisfies ö but not φ
- A satisfies (φ ″ψ) if and only if A satisfies both φ and ψ or satisfies neither one of them

With this definition we can now formalize what it means for a formula ö to be implied by a certain set S of formulas. Informally this is true if in all worlds that are possible given the set of formulas S the formula ö also holds. This leads to the following formal definition: We say that a set S of well-formed formulas semantically entails (or implies) a certain well-formed formula ö if all truth assignments that satisfy all the formulas in S also satisfy φ.

Finally we define syntactical entailment such that ö is syntactically entailed by S if and only if we can derive it with the inference rules that were presented above in a finite number of steps. This allows us to formulate exactly what it means for the set of inference rules to be sound and complete:

Soundness: If the set of well-formed formulas S syntactically entails the well-formed formula ö then S semantically entails ö.

Completeness: If the set of well-formed formulas S semantically entails the well-formed formula ö then S syntactically entails ö.

For the above set of rules this is indeed the case.

Sketch of a soundness proof

Notational conventions: Let G be a variable ranging over sets of sentences. Let A, B and C range over sentences. For "G syntactically entails A" we write "G proves A". For "G semantically entails A" we write "G implies A".

We want to show: (A)(G) (if G proves A, then G implies A).

We note that "G proves A" has an inductive definition, and that gives us the immediate resources for demonstrating claims of the form "If G proves A, then ...". So our proof proceeds by induction.

I. Basis. Show: If A is a member of G, then G implies A.

II. Basis. Show: If A is an axiom, then G implies A.

III. Inductive step (induction on n, the length of the proof):

a. Assume for arbitrary G and A that if G proves A in n or fewer steps, then G implies A.

b. For each possible application of a rule of inference at step n + 1, leading to a new theorem B, show that G implies B.

Notice that Basis Step II can be omitted for natural deduction systems because they have no axioms. When used, Step II involves showing that each of the axioms is a (semantic) logical truth.

The Basis steps demonstrate that the simplest provable sentences from G are also implied by G, for any G. (The proof is simple, since the semantic fact that a set implies any of its members, is also trivial.) The Inductive step will systematically cover all the further sentences that might be provable—by considering each case where we might reach a logical conclusion using an inference rule—and shows that if a new sentence is provable, it is also logically implied. (For example, we might have a rule telling us that from "A" we can derive "A or B". In III.a We assume that if A is provable it is implied. We also know that if A is provable then "A or B" is provable. We have to show that then "A or B" too is implied. We do so by appeal to the semantic definition and the assumption we just made. A is provable from G, we assume. So it is also implied by G. So any semantic valuation making all of G true makes A true. But any valuation making A true makes "A or B" true, by the defined semantics for "or". So any valuation which makes all of G true makes "A or B" true. So "A or B" is implied.) Generally, the Inductive step will consist of a lengthy but simple case-by-case analysis of all the rules of inference, showing that each "preserves" semantic implication.

By the definition of provability, there are no sentences provable other than by being a member of G, an axiom, or following by a rule; so if all of those are semantically implied, the deduction calculus is sound.

Simple and Compound Proposition

A simple proposition is one that does not contain(any other propositions as its part.

A compound proposition is one that is made up of(two or more simple propositions.

We use lower case letters a,b,c... to represent proposition.

Example of simple proposition

It is raining.

This statement can either be TRUE or FALSE.

Example of compound proposition

Today is Sunday and Sunday is a holiday.

This statement contain two simple propositions "Today is Sunday" and "Sunday is a holiday" both the statement can be either TRUE or FALSE.

Operator or Logical Connective

Operator or logical connective are the things that joins simple propositions into compound propositions and joins compound propositions into larger compound propositions.

What is Propositional Logic?

Propositional Logic is a way to represent logic through propositions and logical connectives.

TYPES OF LOGICAL CONNECTIVES (OPERATORS)

Following are the types of logical connectives (operators) used in propositional logic.

- Disjunctive (also called OR)
- Conjunctive (also called AND)
- Conditional (also called Implication)
- Bi-conditional (also called Equivalence)
- Negation (also called NOT)

Disjunctive operator

Disjunctive (also called OR) means one of the two arguments is true or(both of them are true.

We use the word OR and + and (∨ symbols to represent disjunctive.

Example

p + q

p (∨ q

p OR q

They all mean

either p is true, or q is true, or both are true.

Consider two arguments (proposition)

p = Oct 21, 2012 was Sunday

q = Sunday is a holiday

then,

p + q

p (∨ q

p OR q

They all means either p is true, or q is true, or both are true.

i.e., either Oct 21, 2012 was Sunday is true or Sunday is a holiday is true or both are true.

Conjunctive operator

Conjunctive (also called AND) means both the arguments are true. We use the word AND and . & and ∨ symbols to represent conjunctive.

Example

p . q

p & q

p ∨ q

p AND q

They all means both p and q are true.

Consider two arguments (proposition)

p = Oct 21, 2012 was Sunday

q = Sunday is a holiday

then,

p . q

p & q

p ∨ q

p AND q

They all means both p and q are true.

i.e., both Oct 21, 2012 was Sunday and Sunday is a holiday are true.

Conditional (Implication) operator

Conditional also called Implication (If...Then).

Implication means if one argument is true then the other argument is true.

We use the ~ symbol to represent conditional operator.

Example

p ~ q

this means if p is true, then q is true.

Consider two arguments (proposition)

p = 10 is greater than 0

q = 10 is positive

then,

p ~ q

this means if p is true, then q is true.

i.e., if 10 is greater than 0 then 10 is positive.

Bi-conditional (Equivalence) operator

Bi-conditional also called Equivalence (If and only If).

Equivalence means either both arguments are true or both are false.

We use the ~ symbol to represent bi-conditional.

Example

p ~ q

this means

p and q either both are true or both are false.

Consider two arguments (proposition)

p = 10 is greater than 0

q = 10 is positive

then,

p ~ q

this means

p and q either both are true or both are false.

i.e., 10 is greater than 0 and 10 is positive

either both are true or both are false.

Negation (also called NOT)

Negation is an operator that affects only one statement and does not join two statements.

We use the ~ and ′ symbol to represent negation

~p

p′

this means

if p is true then,

~p is false.

Example

Consider the argument (proposition)

p = It is raining

if p is true i.e., it is raining

then,

~p is false i.e., it is not raining.

Graphical calculi

It is possible to generalize the definition of a formal language from a set of finite sequences over a finite basis to include many other sets of mathematical structures, so long as they are built up by finitary means from finite materials. What's more, many of these families of formal structures are especially well-suited for use in logic.

For example, there are many families of graphs that are close enough analogues of formal languages that the concept of a calculus is quite easily and naturally extended to them. Many species of graphs arise as parse graphs in the syntactic analysis of the corresponding families of text structures. The exigencies of practical computation on formal languages frequently demand that text strings be converted into pointer structure renditions of parse graphs, simply as a matter of checking whether strings are well-formed formulas or not.

Once this is done, there are many advantages to be gained from developing the graphical analogue of the calculus on strings. The mapping from strings to parse graphs is called parsing and the inverse mapping from parse graphs to strings is achieved by an operation that is called traversing the graph.

Other logical calculi

Propositional calculus is about the simplest kind of logical calculus in current use. It can be extended in several ways. (Aristotelian "syllogistic" calculus, which is largely supplanted in modern logic, is in some ways simpler – but in other ways more complex – than propositional calculus.) The most immediate way to develop a more complex logical calculus is to introduce rules that are sensitive to more fine-grained details of the sentences being used.

First-order logic (a.k.a. first-order predicate logic) results when the "atomic sentences" of propositional logic are broken up into terms, variables, predicates, and quantifiers, all keeping the rules of propositional logic with some new ones introduced. (For example, from "All dogs are mammals" we may infer "If Rover is a dog then Rover is a mammal".) With the tools of first-order logic it is possible to formulate a number of theories, either with explicit axioms or by rules of inference, that can themselves be treated as logical calculi. Arithmetic is the best known of these; others include set theory and mereology. Second-order logic and other higher-order logics are formal extensions of first-order logic. Thus, it makes sense to refer to propositional logic as "zeroth-order logic", when comparing it with these logics.

Modal logic also offers a variety of inferences that cannot be captured in propositional calculus. For example, from "Necessarily p" we may infer that p. From p we may infer "It is possible that p". The translation between modal logics and algebraic logics concerns classical and intuitionistic logics but with the introduction of a unary operator on Boolean or Heyting algebras, different from the Boolean operations, interpreting the possibility modality, and in the case of Heyting algebra a second operator interpreting necessity (for Boolean algebra this is redundant since necessity is the De Morgan dual of possibility). The first operator preserves 0 and disjunction while the second preserves 1 and conjunction.

Many-valued logics are those allowing sentences to have values other than true and false. (For example, neither and both are standard "extra values"; "continuum logic" allows each sentence to have any of an infinite number of "degrees of truth" between true and false.) These logics often require calculational devices quite distinct from propositional calculus. When the values form a Boolean algebra (which may have more than two or even infinitely many values), many-valued logic reduces to classical logic; many-valued logics are therefore only of independent interest when the values form an algebra that is not Boolean.

3

Boolean Algebras Canonically Defined

Boolean algebra is a mathematically rich branch of abstract algebra. Stanford Encyclopaedia of Philosophy defines Boolean algebra as 'the algebra of two-valued logic with only sentential connectives, or equivalently of algebras of sets under union and complementation.' Just as group theory deals with groups, and linear algebra with vector spaces, Boolean algebras are models of the equational theory of the two values 0 and 1 (whose interpretation need not be numerical). Common to Boolean algebras, groups, and vector spaces is the notion of an algebraic structure, a set closed under some operations satisfying certain equations.

Just as there are basic examples of groups, such as the group $\mathbb{Z}$ of integers and the symmetric group S_n of permutations of n objects, there are also basic examples of Boolean algebras such as the following.

- The algebra of binary digits or bits 0 and 1 under the logical operations including disjunction, conjunction, and negation. Applications include the propositional calculus and the theory of digital circuits.
- The algebra of sets under the set operations including union, intersection, and complement. Applications are far-reaching because set theory is the standard foundations of mathematics.

Boolean algebra thus permits applying the methods of abstract algebra to mathematical logic and digital logic.

Unlike groups of finite order, which exhibit complexity and diversity and whose first-order theory is decidable only in special cases, all finite Boolean algebras share the same theorems and have a decidable first-order theory. Instead, the intricacies of Boolean algebra are divided between the structure of infinite algebras and the algorithmic complexity of their syntactic structure.

DEFINITION

Boolean algebra treats the equational theory of the maximal two-element finitary algebra, called the Boolean prototype, and the models of that theory, called Boolean algebras. These terms are defined as follows.

An algebra is a family of operations on a set, called the underlying set of the algebra. We take the underlying set of the Boolean prototype to be {0,1}.

An algebra is finitary when each of its operations takes only finitely many arguments. For the prototype each argument of an operation is either 0 or 1, as is the result of the operation. The maximal such algebra consists of all finitary operations on {0,1}.

The number of arguments taken by each operation is called the arity of the operation. An operation on {0,1} of arity n, or n-ary operation, can be applied to any of 2^n possible values for its n arguments. For each choice of arguments, the operation may return 0 or 1, whence there are 2^{2n} n-ary operations.

The prototype therefore has two operations taking no arguments, called zeroary or nullary operations, namely zero and one. It has four unary operations, two of which are constant operations, another is the identity, and the most commonly used one, called negation, returns the opposite of its argument: 1 if 0, 0 if 1. It has sixteen binary operations; again two of these are constant, another returns its first argument, yet another returns its second, one is called conjunction and returns 1 if both arguments are 1 and otherwise 0, another is called disjunction and returns 0 if both arguments are 0 and otherwise 1, and so on. The number of (n+1)-ary operations in the prototype is the square of the number of n-ary operations, so there are $16^2 = 256$ ternary operations, $256^2 = 65{,}536$ quaternary operations, and so on.

A family is indexed by an index set. In the case of a family of operations forming an algebra, the indices are called operation symbols, constituting the language of that algebra. The operation indexed by each symbol is called the denotation or interpretation of that symbol. Each operation symbol specifies the arity of its interpretation, whence all possible interpretations of a symbol have the same arity. In general it is possible for an algebra to interpret distinct symbols with the same operation, but this is not the case for the prototype, whose symbols are in one-one correspondence with its operations. The prototype therefore has 2^{2n} n-ary operation symbols, called the Boolean operation symbols and forming the language of Boolean algebra. Only a few operations have conventional symbols, such as ¬ for negation, ‘‘‘ for conjunction, and (“ for disjunction. It is convenient to consider the i-th n-ary symbol to be ${}^nf_i$ as done below in the section on truth tables.

An equational theory in a given language consists of equations between terms built up from variables using symbols of that language. Typical equations in the language of Boolean algebra are $x \wedge y = y \wedge x$, $x \wedge x = x$, $x \wedge \neg x = y \wedge \neg y$, and $x \wedge y = x$.

An algebra satisfies an equation when the equation holds for all possible values of its variables in that algebra when the operation symbols are interpreted as specified by that algebra. The laws of Boolean algebra are the equations in the language of Boolean algebra satisfied by the prototype. The first three of the above examples are Boolean laws, but not the fourth since 1'"0 '" 1.

The equational theory of an algebra is the set of all equations satisfied by the algebra. The laws of Boolean algebra therefore constitute the equational theory of the Boolean prototype.

A model of a theory is an algebra interpreting the operation symbols in the language of the theory and satisfying the equations of the theory.

A Boolean algebra is any model of the laws of Boolean algebra.

That is, a Boolean algebra is a set and a family of operations thereon interpreting the Boolean operation symbols and satisfying the same laws as the Boolean prototype.

If we define a homologue of an algebra to be a model of the equational theory of that algebra, then a Boolean algebra can be defined as any homologue of the prototype.

Example 1. The Boolean prototype is a Boolean algebra, since trivially it satisfies its own laws. It is thus the prototypical Boolean algebra. We did not call it that initially in order to avoid any appearance of circularity in the definition.

Basis

The operations need not be all explicitly stated. A basis is any set from which the remaining operations can be obtained by composition. A "Boolean algebra" may be defined from any of several different bases. Three bases for Boolean algebra are in common use, the lattice basis, the ring basis, and the Sheffer stroke or NAND basis. These bases impart respectively a logical, an arithmetical, and a parsimonious character to the subject.

- The lattice basis originated in the 19th century with the work of Boole, Peirce, and others seeking an algebraic formalization of logical thought processes.
- The ring basis emerged in the 20th century with the work of Zhegalkin and Stone and became the basis of choice for algebraists coming to the subject from a background in abstract algebra. Most treatments of Boolean algebra assume the lattice basis, a notable exception being

Halmos[1963] whose linear algebra background evidently endeared the ring basis to him.

- Since all finitary operations on {0,1} can be defined in terms of the Sheffer stroke NAND (or its dual NOR), the resulting economical basis has become the basis of choice for analyzing digital circuits, in particular gate arrays in digital electronics.

The common elements of the lattice and ring bases are the constants 0 and 1, and an associative commutative binary operation, called meet x′dy in the lattice basis, and multiplication xy in the ring basis. The distinction is only terminological. The lattice basis has the further operations of join, x(dy, and complement, ¬x. The ring basis has instead the arithmetic operation x•dy of addition (the symbol •d is used in preference to + because the latter is sometimes given the Boolean reading of join).

To be a basis is to yield all other operations by composition, whence any two bases must be intertranslatable. The lattice basis translates x(dy to the ring basis as x•dy•dxy, and ¬x as x•d1. Conversely the ring basis translates x•dy to the lattice basis as (x(dy)′d¬(x′dy).

Both of these bases allow Boolean algebras to be defined via a subset of the equational properties of the Boolean operations. For the lattice basis, it suffices to define a Boolean algebra as a distributive lattice satisfying x′d¬x = 0 and x(d¬x = 1, called a complemented distributive lattice. The ring basis turns a Boolean algebra into a Boolean ring, namely a ring satisfying $x^2 = x$.

Emil Post gave a necessary and sufficient condition for a set of operations to be a basis for the nonzeroary Boolean operations. A nontrivial property is one shared by some but not all operations making up a basis. Post listed five nontrivial properties of operations, identifiable with the five Post's classes, each preserved by composition, and showed that a set of operations formed a basis if, for each property, the set contained an operation lacking that property. (The converse of Post's theorem, extending "if" to "if and only if," is the easy observation that a property from among these five holding of every operation in a candidate basis will also hold of every operation formed by composition from that candidate, whence by nontriviality of that property the candidate will fail to be a basis.) Post's five properties are:

- monotone, no 0-1 input transition can cause a 1-0 output transition;
- affine, representable with Zhegalkin polynomials that lack bilinear or higher terms, e.g. x⊕y⊕1 but not xy;
- self-dual, so that complementing all inputs complements the output, as with x, or the median operator xy⊕yz⊕zx, or their negations;
- strict (mapping the all-zeros input to zero);

- costrict (mapping all-ones to one).

The NAND (dually NOR) operation lacks all these, thus forming a basis by itself.

Truth tables

The finitary operations on {0,1} may be exhibited as truth tables, thinking of 0 and 1 as the truth values false and true. They can be laid out in a uniform and application-independent way that allows us to name, or at least number, them individually. These names provide a convenient shorthand for the Boolean operations. The names of the n-ary operations are binary numbers of 2^n bits. There being 2^{2n} such operations, one cannot ask for a more succinct nomenclature. Note that each finitary operation can be called a switching function.

Examples

Bit vectors

Example 2. All bit vectors of a given length form a Boolean algebra "pointwise", meaning that any n-ary Boolean operation can be applied to n bit vectors one bit position at a time. For example, the ternary OR of three bit vectors each of length 4 is the bit vector of length 4 formed by oring the three bits in each of the four bit positions, thus 0100("1000("1001 = 1101. Another example is the truth tables above for the n-ary operations, whose columns are all the bit vectors of length 2^n and which therefore can be combined pointwise whence the n-ary operations form a Boolean algebra. This works equally well for bit vectors of finite and infinite length, the only rule being that the bit positions all be indexed by the same set in order that "corresponding position" be well defined.

The atoms of such an algebra are the bit vectors containing exactly one 1. In general the atoms of a Boolean algebra are those elements x such that x∧y has only two possible values, x or 0.

Power set algebra

Example 3. The power set algebra, the set 2^W of all subsets of a given set W. This is just Example 2 in disguise, with W serving to index the bit positions. Any subset X of W can be viewed as the bit vector having 1's in just those bit positions indexed by elements of X. Thus the all-zero vector is the empty subset of W while the all-ones vector is W itself, these being the constants 0 and 1 respectively of the power set algebra. The counterpart of disjunction x("y is union X*"Y, while that of conjunction x∧y is intersection X)"Y. Negation ¬x becomes ~X, complement relative to W. There is also set difference X"Y = X)"~Y, symmetric difference (X"Y)*"(Y"X), ternary union X*"Y*"Z, and so on. The atoms here are the singletons, those subsets with exactly one element.

Examples 2 and 3 are special cases of a general construct of algebra called direct product, applicable not just to Boolean algebras but all kinds of algebra including groups, rings, etc. The direct product of any family B_i of Boolean algebras where i ranges over some index set I (not necessarily finite or even countable) is a Boolean algebra consisting of all I-tuples $(\ldots x_i,\ldots)$ whose i-th element is taken from B_i. The operations of a direct product are the corresponding operations of the constituent algebras acting within their respective coordinates; in particular operation ${}^nf_j$ of the product operates on n I-tuples by applying operation ${}^nf_j$ of B_i to the n elements in the i-th coordinate of the n tuples, for all i in I.

When all the algebras being multiplied together in this way are the same algebra A we call the direct product a direct power of A. The Boolean algebra of all 32-bit bit vectors is the two-element Boolean algebra raised to the 32nd power, or power set algebra of a 32-element set, denoted 2^{32}. The Boolean algebra of all sets of integers is 2^Z. All Boolean algebras we have exhibited thus far have been direct powers of the two-element Boolean algebra, justifying the name "power set algebra".

REPRESENTATION THEOREMS

It can be shown that every finite Boolean algebra is isomorphic to some power set algebra. Hence the cardinality (number of elements) of a finite Boolean algebra is a power of 2, namely one of $1,2,4,8,\ldots,2^n,\ldots$ This is called a representation theorem as it gives insight into the nature of finite Boolean algebras by giving a representation of them as power set algebras.

This representation theorem does not extend to infinite Boolean algebras: although every power set algebra is a Boolean algebra, not every Boolean algebra need be isomorphic to a power set algebra. In particular, whereas there can be no countably infinite power set algebras (the smallest infinite power set algebra is the power set algebra 2^N of sets of natural numbers, shown by Cantor to be uncountable), there exist various countably infinite Boolean algebras.

To go beyond power set algebras we need another construct. A subalgebra of an algebra A is any subset of A closed under the operations of A. Every subalgebra of a Boolean algebra A must still satisfy the equations holding of A, since any violation would constitute a violation for A itself. Hence every subalgebra of a Boolean algebra is a Boolean algebra.

A subalgebra of a power set algebra is called a field of sets; equivalently a field of sets is a set of subsets of some set W including the empty set and W and closed under finite union and complement with respect to W (and hence also under finite intersection). Birkhoff's [1935] representation theorem for Boolean algebras states that every Boolean algebra is isomorphic to a field of sets. Now Birkhoff's HSP theorem for varieties can be stated as, every class of models of the equational theory of a class C of algebras is the Homomorphic image of a Subalgebra of a direct Product of algebras of C. Normally all three of H, S, and P are needed; what the first of these two Birkhoff theorems shows is that for the special case of the variety of Boolean algebras Homomorphism can be replaced by Isomorphism. Birkhoff's HSP theorem for varieties in general therefore becomes Birkhoff's ISP theorem for the variety of Boolean algebras.

Other examples

It is convenient when talking about a set X of natural numbers to view it as a sequence $x_0,x_1,x_2,...$ of bits, with $x_i = 1$ if and only if i " X. This viewpoint will make it easier to talk about subalgebras of the power set algebra 2^N, which this viewpoint makes the Boolean algebra of all sequences of bits. It also fits well with the columns of a truth table: when a column is read from top to bottom it constitutes a sequence of bits, but at the same time it can be viewed as the set of those valuations (assignments to variables in the left half of the table) at which the function represented by that column evaluates to 1.

Example 4. Ultimately constant sequences. Any Boolean combination of ultimately constant sequences is ultimately constant; hence these form a Boolean algebra. We can identify these with the integers by viewing the ultimately-zero sequences as nonnegative binary numerals (bit 0 of the sequence being the low-order bit) and the ultimately-one sequences as negative binary numerals (think two's complement arithmetic with the all-ones sequence being "1). This makes the integers a Boolean algebra, with union being bit-wise OR and complement being "x"1. There are only countably many integers, so this infinite Boolean algebra is countable. The atoms are the powers of two, namely 1,2,4,.... Another way of describing this algebra is as the set of all finite and cofinite sets of natural numbers, with the ultimately all-ones sequences corresponding to the cofinite sets, those sets omitting only finitely many natural numbers.

Example 5. Periodic sequence. A sequence is called periodic when there exists some number n > 0, called a witness to periodicity, such that $x_i = x_{i+n}$ for all i e" 0. The period of a periodic sequence is its least witness. Negation leaves period unchanged, while the disjunction of two periodic sequences is periodic, with period at most the least common multiple of the periods of the two arguments (the period can be as small as 1, as happens with the union of any sequence and its complement). Hence the periodic sequences form a Boolean algebra.

Example 5 resembles Example 4 in being countable, but differs in being atomless. The latter is because the conjunction of any nonzero periodic sequence x with a sequence of greater period is neither 0 nor x. It can be shown that all countably infinite atomless Boolean algebras are isomorphic, that is, up to isomorphism there is only one such algebra.

Example 6. Periodic sequence with period a power of two. This is a proper subalgebra of Example 5 (a proper subalgebra equals the intersection of itself with its algebra). These can be understood as the finitary operations, with the first period of such a sequence giving the truth table of the operation it represents. For example, the truth table of x_0 in the table of binary operations, namely $^2f_{10}$, has period 2 (and so can be recognized as using only the first variable) even though 12 of the binary operations have period 4. When the period is 2^n the operation only depends on the first n variables, the sense in which the operation is finitary. This example is also a countably infinite atomless Boolean algebra. Hence Example 5 is isomorphic to a proper subalgebra of itself! Example 6, and hence Example 5, constitutes the free Boolean algebra on countably many generators, meaning the Boolean algebra of all finitary operations on a countably infinite set of generators or variables.

Example 7. Ultimately periodic sequences, sequences that become periodic after an initial finite bout of lawlessness. They constitute a proper extension of Example 5 (meaning that Example 5 is a proper subalgebra of Example 7) and also of Example 4, since constant sequences are periodic with period one. Sequences may vary as to when they settle down, but any finite set of sequences will all eventually settle down no later than their slowest-to-settle member, whence ultimately periodic sequences are closed under all Boolean operations and so form a Boolean algebra. This example has the same atoms and coatoms as Example 4, whence it is not atomless and therefore not isomorphic to Example 5/6. However it contains an infinite atomless subalgebra, namely Example 5, and so is not isomorphic to Example 4, every subalgebra of which must be a Boolean algebra of finite sets and their complements and therefore atomic. This example is isomorphic to the direct product of Examples 4 and 5, furnishing another description of it.

Example 8. The direct product of a Periodic Sequence (Example 5) with any finite but nontrivial Boolean algebra. (The trivial one-element Boolean algebra is the unique finite atomless Boolean algebra.) This resembles Example 7 in having both atoms and an atomless subalgebra, but differs in having only finitely many atoms. Example 8 is in fact an infinite family of examples, one for each possible finite number of atoms.

These examples by no means exhaust the possible Boolean algebras, even the countable ones. Indeed, there are uncountably many nonisomorphic countable Boolean algebras, which Jussi Ketonen [1978] classified completely in terms of invariants representable by certain hereditarily countable sets.

BOOLEAN ALGEBRAS OF BOOLEAN OPERATIONS

The n-ary Boolean operations themselves constitute a power set algebra 2^W, namely when W is taken to be the set of 2^n valuations of the n inputs. In terms of the naming system of operations ${}^nf_i$ where i in binary is a column of a truth table, the columns can be combined with Boolean operations of any arity to produce other columns present in the table. That is, we can apply any Boolean operation of arity m to m Boolean operations of arity n to yield a Boolean operation of arity n, for any m and n.

The practical significance of this convention for both software and hardware is that n-ary Boolean operations can be represented as words of the appropriate length. For example, each of the 256 ternary Boolean operations can be represented as an unsigned byte. The available logical operations such as AND and OR can then be used to form new operations. If we take x, y, and z (dispensing with subscripted variables for now) to be 10101010, 11001100, and 11110000 respectively (170, 204, and 240 in decimal, 0xaa, 0xcc, and 0xf0 in hexadecimal), their pairwise conjunctions are $x \wedge y$ = 10001000, $y \wedge z$ = 11000000, and $z \wedge x$ = 10100000, while their pairwise disjunctions are x(“y = 11101110, y(“z = 11111100, and z(“x = 11111010. The disjunction of the three conjunctions is 11101000, which also happens to be the conjunction of three disjunctions. We have thus calculated, with a dozen or so logical operations on bytes, that the two ternary operations

$$(x \wedge y) \vee (y \wedge z) \vee (z \wedge x)$$

and

$$(x \vee y) \wedge (y \vee z) \wedge (z \vee x)$$

are actually the same operation. That is, we have proved the equational identity

$$(x \wedge y) \vee (y \wedge z) \vee (z \wedge x) = (x \vee y) \wedge (y \vee z) \wedge (z \vee x),$$

for the two-element Boolean algebra. By the definition of "Boolean algebra" this identity must therefore hold in every Boolean algebra.

This ternary operation incidentally formed the basis for Grau's [1947] ternary Boolean algebras, which he axiomatized in terms of this operation and negation. The operation is symmetric, meaning that its value is independent of any of the 3! = 6 permutations of its arguments. The two halves of its truth table 11101000 are the truth tables for (", 1110, and '", 1000, so the operation can be phrased as if z then x("y else x∧y. Since it is symmetric it can equally well be phrased as either of if x then y("z else y∧z, or if y then z("x else z∧x. Viewed as a labeling of the 8-vertex 3-cube, the upper half is labeled 1 and the lower half 0; for this reason it has been called the median operator, with the evident generalization to any odd number of variables (odd in order to avoid the tie when exactly half the variables are 0).

AXIOMATIZING BOOLEAN ALGEBRAS

The technique we just used to prove an identity of Boolean algebra can be generalized to all identities in a systematic way that can be taken as a sound and complete axiomatization of, or axiomatic system for, the equational laws of Boolean logic. The customary formulation of an axiom system consists of a set of axioms that "prime the pump" with some initial identities, along with a set of inference rules for inferring the remaining identities from the axioms and previously proved identities. In principle it is desirable to have finitely many axioms; however as a practical matter it is not necessary since it is just as effective to have a finite axiom schema having infinitely many instances each of which when used in a proof can readily be verified to be a legal instance, the approach we follow here.

Boolean identities are assertions of the form s = t where s and t are n-ary terms, by which we shall mean here terms whose variables are limited to x_0 through x_{n-1}. An n-ary term is either an atom or an application. An application ${}^{m}f_i(t_0,\ldots,t_{m-1})$ is a pair consisting of an m-ary operation ${}^{m}f_i$ and a list or m-tuple $(t_0,\ldots,t_{m-1})$ of m n-ary terms called operands.

Associated with every term is a natural number called its height. Atoms are of zero height, while applications are of height one plus the height of their highest operand.

Now what is an atom? Conventionally an atom is either a constant (0 or 1) or a variable x_i where $0 \leq i < n$. For the proof technique here it is convenient to define atoms instead to be n-ary operations ${}^{n}f_i$, which although treated here as atoms nevertheless mean the same as ordinary terms of the exact form ${}^{n}f_i(x_0,\ldots,x_{n-1})$ (exact in that the variables must listed in the order shown without repetition or omission). This is not a restriction because atoms of this form include all the ordinary atoms, namely the constants 0 and 1, which arise here as the n-ary operations ${}^{n}f_0$ and ${}^{n}f_{''1}$ for each n (abbreviating 2^{2n}"1 to "1), and the variables $x_0,\ldots,x_{n-1}$ as can be seen from the truth tables where x_0 appears as both the unary operation ${}^{1}f_2$ and the binary operation ${}^{2}f_{10}$ while x_1 appears as ${}^{2}f_{12}$.

The following axiom schema and three inference rules axiomatize the Boolean algebra of n-ary terms.

A1. ${}^mf_i({}^nf_{j0},\ldots,{}^nf_{jm-1}) = {}^nf_{io5}$ where $(io5)_v = i_{5v}$, with 5 being j transpose, defined by $(5_v)_u = (j_u)_v$.

R1. With no premises infer t = t.

R2. From s = u and t = u infer s = t where s, t, and u are n-ary terms.

R3. From $s_0 = t_0,\ldots,s_{m-1} = t_{m-1}$ infer ${}^mf_i(s_0,\ldots,s_{m-1}) = {}^mf_i(t_0,\ldots,t_{m-1})$, where all terms s_i, t_i are n-ary.

The meaning of the side condition on A1 is that io5 is that 2^n-bit number whose v-th bit is the 5_v-th bit of i, where the ranges of each quantity are u: m, v: 2^n, j_u: 2^{2n}, and 5_v: 2^m. (So j is an m-tuple of 2^n-bit numbers while 5 as the transpose of j is a 2^n-tuple of m-bit numbers. Both j and 5 therefore contain $m2^n$ bits.)

A1 is an axiom schema rather than an axiom by virtue of containing metavariables, namely m, i, n, and j_0 through j_{m-1}. The actual axioms of the axiomatization are obtained by setting the metavariables to specific values. For example, if we take $m = n = i = j_0 = 1$, we can compute the two bits of io5 from $i_1 = 0$ and $i_0 = 1$, so io5 = 2 (or 10 when written as a two-bit number). The resulting instance, namely ${}^1f_1({}^1f_1) = {}^1f_2$, expresses the familiar axiom $\neg\neg x = x$ of double negation. Rule R3 then allows us to infer $\neg\neg\neg x = \neg x$ by taking s_0 to be ${}^1f_1({}^1f_1)$ or $\neg\neg x_0$, t_0 to be ${}^1f_2$ or x_0, and ${}^mf_i$ to be ${}^1f_1$ or $\neg$.

For each m and n there are only finitely many axioms instantiating A1, namely $2^{2m} \times (2^{2n})^m$. Each instance is specified by 2^m+m2^n bits.

We treat R1 as an inference rule, even though it is like an axiom in having no premises, because it is a domain-independent rule along with R2 and R3 common to all equational axiomatizations, whether of groups, rings, or any other variety. The only entity specific to Boolean algebras is axiom schema A1. In this way when talking about different equational theories we can push the rules to one side as being independent of the particular theories, and confine attention to the axioms as the only part of the axiom system characterizing the particular equational theory at hand.

This axiomatization is complete, meaning that every Boolean law s = t is provable in this system. One first shows by induction on the height of s that every Boolean law for which t is atomic is provable, using R1 for the base case (since distinct atoms are never equal) and A1 and R3 for the induction step (s an application). This proof strategy amounts to a recursive procedure for evaluating s to yield an atom. Then to prove s = t in the general case when t may be an application, use the fact that if s = t is an identity then s and t must evaluate to the same atom, call it u. So first prove s = u and t = u as above, that is, evaluate s and t using A1, R1, and R3, and then invoke R2 to infer s = t.

In A1, if we view the number n^m as the function type m→n, and m_n as the application m(n), we can reinterpret the numbers i, j, 5, and io5 as functions of type i: (m→2)→2, j: m→((n→2)→2), 5: (n→2)→(m→2), and io5: (n→2)→2. The definition $(io5)_v = i_{5v}$ in A1 then translates to (io5)(v) = i(5(v)), that is, io5 is defined to be composition of i and 5 understood as functions. So the content of A1 amounts to defining term application to be essentially composition, modulo the need to transpose the m-tuple j to make the types match up suitably for composition. This composition is the one in Lawvere's previously mentioned category of power sets and their functions. In this way we have translated the commuting diagrams of that category, as the equational theory of Boolean algebras, into the equational consequences of A1 as the logical representation of that particular composition law.

UNDERLYING LATTICE STRUCTURE

Underlying every Boolean algebra B is a partially ordered set or poset (B,≤). The partial order relation is defined by x ≤ y just when x = x∧y, or equivalently when y = x(“y. Given a set X of elements of a Boolean algebra, an upper bound on X is an element y such that for every element x of X, x ≤ y, while a lower bound on X is an element y such that for every element x of X, y ≤ x.

A sup of X is a least upper bound on X, namely an upper bound on X that is less or equal to every upper bound on X. Dually an (inf) of X is a greatest lower bound on X. The sup of x and y always exists in the underlying poset of a Boolean algebra, being x(“y, and likewise their inf exists, namely x∧y. The empty sup is 0 (the bottom element) and the empty inf is 1 (top). It follows that every finite set has both a sup and an inf. Infinite subsets of a Boolean algebra may or may not have a sup and/or an inf; in a power set algebra they always do.

Any poset (B,≤) such that every pair x,y of elements has both a sup and an inf is called a lattice. We write x(“y for the sup and x∧y for the inf. The underlying poset of a Boolean algebra always forms a lattice. The lattice is said to be distributive when x∧(y(“z) = (x∧y)(“(x∧z), or equivalently when x(“(y∧z) = (x(“y)∧(x(“z), since either law implies the other in a lattice. These are laws of Boolean algebra whence the underlying poset of a Boolean algebra forms a distributive lattice.

Given a lattice with a bottom element 0 and a top element 1, a pair x,y of elements is called complementary when x∧y = 0 and x(“y = 1, and we then say that y is a complement of x and vice versa. Any element x of a distributive lattice with top and bottom can have at most one complement. When every element of a lattice has a complement the lattice is called complemented. It follows that in a complemented distributive lattice, the complement of an element always exists and is unique, making complement a unary operation. Furthermore, every complemented distributive lattice forms a Boolean algebra, and conversely every Boolean algebra forms a complemented distributive lattice. This provides an alternative definition of a Boolean algebra, namely as any complemented distributive lattice. Each of these three properties can be axiomatized with finitely many equations, whence these equations taken together constitute a finite axiomatization of the equational theory of Boolean algebras.

In a class of algebras defined as all the models of a set of equations, it is usually the case that some algebras of the class satisfy more equations than just those needed to qualify them for the class. The class of Boolean algebras is unusual in that, with a single exception, every Boolean algebra satisfies exactly the Boolean identities and no more. The exception is the one-element Boolean algebra, which necessarily satisfies every equation, even x = y, and is therefore sometimes referred to as the inconsistent Boolean algebra.

Boolean homomorphisms

A Boolean homomorphism is a function h: A→B between Boolean algebras A,B such that for every Boolean operation ${}^{m}f_i$:

$$h({}^{m}f_i(x_0,\ldots,x_{m-1})) = {}^{m}f_i(h(x_0,\ldots,x_{m-1}))$$

The category Bool of Boolean algebras has as objects all Boolean algebras and as morphisms the Boolean homomorphisms between them.

There exists a unique homomorphism from the two-element Boolean algebra 2 to every Boolean algebra, since homomorphisms must preserve the two constants and those are the only elements of 2. A Boolean algebra with this property is called an initial Boolean algebra. It can be shown that any two initial Boolean algebras are isomorphic, so up to isomorphism 2 is the initial Boolean algebra.

In the other direction, there may exist many homomorphisms from a Boolean algebra B to 2. Any such homomorphism partitions B into those elements mapped to 1 and those to 0. The subset of B consisting of the former is called an ultrafilter of B. When B is finite its ultrafilters pair up with its atoms; one atom is mapped to 1 and the rest to 0. Each ultrafilter of B thus consists of an atom of B and all the elements above it; hence exactly half the elements of B are in the ultrafilter, and there as many ultrafilters as atoms.

For infinite Boolean algebras the notion of ultrafilter becomes considerably more delicate. The elements greater or equal than an atom always form an ultrafilter but so do many other sets; for example in the Boolean algebra of finite and cofinite sets of integers the cofinite sets form an ultrafilter even though none of them are atoms. Likewise the powerset of the integers has among its ultrafilters the set of all subsets containing a given integer; there are countably many of these "standar≤ ultrafilters, which may be identified with the integers themselves, but there are uncountably many more "nonstandar≤ ultrafilters. These form the basis for nonstandard analysis, providing representations for such classically inconsistent objects as infinitesimals and delta functions.

INFINITARY EXTENSIONS

Recall the definition of sup and inf from the section above on the underlying partial order of a Boolean algebra. A complete Boolean algebra is one every subset of which has both a sup and an inf, even the infinite subsets. Gaifman [1964] and Hales [1964] independently showed that infinite free complete Boolean algebras do not exist. This suggests that a logic with set-sized-infinitary operations may have class-many terms—just as a logic with finitary operations may have infinitely many terms.

There is however another approach to introducing infinitary Boolean operations: simply drop "finitary" from the definition of Boolean algebra. A model of the equational theory of the algebra of all operations on {0,1} of arity up to the cardinality of the model is called a complete atomic Boolean algebra, or CABA. (In place of this awkward restriction on arity we could allow any arity, leading to a different awkwardness, that the signature would then be larger than any set, that is, a proper class. One benefit of the latter approach is that it simplifies the definition of homomorphism between CABAs of different cardinality.) Such an algebra can be defined equivalently as a complete Boolean algebra that is atomic, meaning that every element is a sup of some set of atoms. Free CABAs exist for all cardinalities of a set V of generators, namely the power set algebra 2^{2V}, this being the obvious generalization of the finite free Boolean algebras. This neatly rescues infinitary Boolean logic from the fate the Gaifman–Hales result seemed to consign it to.

The nonexistence of free complete Boolean algebras can be traced to failure to extend the equations of Boolean logic suitably to all laws that should hold for infinitary conjunction and disjunction, in particular the neglect of distributivity in the definition of complete Boolean algebra. A complete Boolean algebra is called completely distributive when arbitrary conjunctions distribute over arbitrary disjunctions and vice versa. A Boolean algebra is a CABA if and only if it is complete and completely distributive, giving a third definition of CABA. A fourth definition is as any Boolean algebra isomorphic to a power set algebra.

A complete homomorphism is one that preserves all sups that exist, not just the finite sups, and likewise for infs. The category CABA of all CABAs and their complete homomorphisms is dual to the category of sets and their functions, meaning that it is equivalent to the opposite of that category (the category resulting from reversing all morphisms). Things are not so simple for the category Bool of Boolean algebras and their homomorphisms, which Marshall Stone showed in effect (though he lacked both the language and the conceptual framework to make the duality explicit) to be dual to the category of totally disconnected compact Hausdorff spaces, subsequently called Stone spaces.

Another infinitary class intermediate between Boolean algebras and complete Boolean algebras is the notion of a sigma-algebra. This is defined analogously to complete Boolean algebras, but with sups and infs limited to countable arity. That is, a sigma-algebra is a Boolean algebra with all countable sups and infs. Because the sups and infs are of bounded cardinality, unlike the situation with complete Boolean algebras, the Gaifman-Hales result does not apply and free sigma-algebras do exist. Unlike the situation with CABAs however, the free countably generated sigma algebra is not a power set algebra.

OTHER DEFINITIONS OF BOOLEAN ALGEBRA

We have already encountered several definitions of Boolean algebra, as a model of the equational theory of the two-element algebra, as a complemented distributive lattice, as a Boolean ring, and as a product-preserving functor from a certain category (Lawvere). Two more definitions worth mentioning are:

Stone (1936)

A Boolean algebra is the set of all clopen sets of a topological space. It is no limitation to require the space to be a totally disconnected compact Hausdorff space, or Stone space, that is, every Boolean algebra arises in this way, up to isomorphism. Moreover, if the two Boolean algebras formed as the clopen sets of two Stone spaces are isomorphic, so are the Stone spaces themselves, which is not the case for arbitrary topological spaces. This is just the reverse direction of the duality mentioned earlier from Boolean algebras to Stone spaces. This definition is fleshed out by the next definition.

Johnstone (1982)

A Boolean algebra is a filtered colimit of finite Boolean algebras.

(The circularity in this definition can be removed by replacing "finite Boolean algebra" by "finite power set" equipped with the Boolean operations standardly interpreted for power sets.) To put this in perspective, infinite sets arise as filtered colimits of finite sets, infinite CABAs as filtered limits of finite power set algebras, and infinite Stone spaces as filtered limits of finite sets. Thus if one starts with the finite sets and asks how these generalize to infinite objects, there are two ways: "adding" them gives ordinary or inductive sets while "multiplying" them gives Stone spaces or profinite sets. The same choice exists for finite power set algebras as the duals of finite sets: addition yields Boolean algebras as inductive objects while multiplication yields CABAs or power set algebras as profinite objects.

A characteristic distinguishing feature is that the underlying topology of objects so constructed, when defined so as to be Hausdorff, is discrete for inductive objects and compact for profinite objects. The topology of finite Hausdorff spaces is always both discrete and compact, whereas for infinite spaces "discrete"' and "compact" are mutually exclusive. Thus when generalizing finite algebras (of any kind, not just Boolean) to infinite ones, "discrete" and "compact" part company, and one must choose which one to retain. The general rule, for both finite and infinite algebras, is that finitary algebras are discrete, whereas their duals are compact and feature infinitary operations. Between these two extremes, there are many intermediate infinite Boolean algebras whose topology is neither discrete nor compact.

Canonical normal form

In Boolean algebra, any Boolean function can be put into the canonical disjunctive normal form (CDNF) or minterm canonical form and its dual canonical conjunctive normal form (CCNF) or maxterm canonical form. Other canonical forms include the complete sum of prime implicants or Blake canonical form (and its dual), and the algebraic normal form (also called Zhegalkin or Reed–Muller).

Minterms are called products because they are the logical AND of a set of variables, and maxterms are called sums because they are the logical OR of a set of variables. These concepts are dual because of their complementary-symmetry relationship as expressed by De Morgan's laws.

Two dual canonical forms of any Boolean function are a "sum of minterms" and a "product of maxterms." The term "Sum of Products" (SoP or SOP) is widely used for the canonical form that is a disjunction (OR) of minterms. Its De Morgan dual is a "Product of Sums" (PoS or POS) for the canonical form that is a conjunction (AND) of maxterms. These forms can be useful for the simplification of these functions, which is of great importance in the optimization of Boolean formulas in general and digital circuits in particular.

APPLICATION EXAMPLE

The sample truth tables for minterms and maxterms above are sufficient to establish the canonical form for a single bit position in the addition of binary numbers, but are not sufficient to design the digital logic unless your inventory of gates includes AND and OR. Where performance is an issue (as in the Apollo Guidance Computer), the available parts are more likely to be NAND and NOR because of the complementing action inherent in transistor logic. The values are defined as voltage states, one near ground and one near the DC supply voltage V_{cc}, e.g. +5 VDC. If the higher voltage is defined as the 1 "true" value, a NOR gate is the simplest possible useful logical element.

Specifically, a 3-input NOR gate may consist of 3 bipolar junction transistors with their emitters all grounded, their collectors tied together and linked to V_{cc} through a load impedance. Each base is connected to an input signal, and the common collector point presents the output signal. Any input that is a 1 (high voltage) to its base shorts its transistor's emitter to its collector, causing current to flow through the load impedance, which brings the collector voltage (the output) very near to ground. That result is independent of the other inputs. Only when all 3 input signals are 0 (low voltage) do the emitter-collector impedances of all 3 transistors remain very high. Then very little current flows, and the voltage-divider effect with the load impedance imposes on the collector point a high voltage very near to V_{cc}.

The complementing property of these gate circuits may seem like a drawback when trying to implement a function in canonical form, but there is a compensating bonus: such a gate with only one input implements the complementing function, which is required frequently in digital logic.

This example assumes the Apollo parts inventory: 3-input NOR gates only, but the discussion is simplified by supposing that 4-input NOR gates are also available (in Apollo, those were compounded out of pairs of 3-input NORs).

Canonical and non-canonical consequences of NOR gates

A set of 8 NOR gates, if their inputs are all combinations of the direct and complement forms of the 3 input variables ci, x, and y, always produce minterms, never maxterms—that is, of the 8 gates required to process all combinations of 3 input variables, only one has the output value 1. That's because a NOR gate, despite its name, could better be viewed (using De Morgan's law) as the AND of the complements of its input signals.

The reason this is not a problem is the duality of minterms and maxterms, i.e. each maxterm is the complement of the like-indexed minterm, and vice versa. In the minterm example above, we wrote u (c_i, x, y) = m_1 + m_2 + m_4 + m_7 but to perform this with a 4-input NOR gate we need to restate it as a product of sums (PoS), where the sums are the opposite maxterms. That is,

$$u(ci, x, y) = \text{AND}(M_0, M_3, M_5, M_6) = \text{NOR}(m_0, m_3, m_5, m_6).$$

Truth tables

ci	x	y	M_0	M_3	M_5	M_6	AND	u(ci,x,y)
0	0	0	0	1	1	1	0	0
0	0	1	1	1	1	1	1	1
0	1	0	1	1	1	1	1	1
0	1	1	1	0	1	1	0	0
1	0	0	1	1	1	1	1	1
1	0	1	1	1	0	1	0	0
1	1	0	1	1	1	0	0	0
1	1	1	1	1	1	1	1	1

ci	x	y	m_0	m_3	m_5	m_6	NOR	u(ci,x,y)
0	0	0	1	0	0	0	0	0
0	0	1	0	0	0	0	1	1
0	1	0	0	0	0	0	1	1
0	1	1	0	1	0	0	0	0
1	0	0	0	0	0	0	1	1
1	0	1	0	0	1	0	0	0
1	1	0	0	0	0	1	0	0
1	1	1	0	0	0	0	1	1

In the maxterm example above, we wrote $co(ci, x, y) = M_0M_1M_2M_4$ but to perform this with a 4-input NOR gate we need to notice the equality to the NOR of the same minterms. That is,

$$co(ci,x,y) = \text{AND}(M_0, M_1, M_2, M_4) = \text{NOR}(m_0, m_1, m_2, m_4).$$

Truth tables

ci	x	y	M₀	M₁	M₂	M₄	AND	co(ci,x,y)
0	0	0	0	1	1	1	0	0
0	0	1	1	0	1	1	0	0
0	1	0	1	1	0	1	0	0
0	1	1	1	1	1	1	1	1
1	0	0	1	1	1	0	0	0
1	0	1	1	1	1	1	1	1
1	1	0	1	1	1	1	1	1
1	1	1	1	1	1	1	1	1
ci	**x**	**y**	**m₀**	**m₁**	**m₂**	**m₄**	**NOR**	**co(ci,x,y)**
0	0	0	1	0	0	0	0	0
0	0	1	0	1	0	0	0	0
0	1	0	0	0	1	0	0	0
0	1	1	0	0	0	0	1	1
1	0	0	0	0	0	1	0	0
1	0	1	0	0	0	0	1	1
1	1	0	0	0	0	0	1	1
1	1	1	0	0	0	0	1	1

Design trade-offs considered in addition to canonical forms

One might suppose that the work of designing an adder stage is now complete, but we haven't addressed the fact that all 3 of the input variables have to appear in both their direct and complement forms. There's no difficulty about the addends x and y in this respect, because they are static throughout the addition and thus are normally held in latch circuits that routinely have both direct and complement outputs. (The simplest latch circuit made of NOR gates is a pair of gates cross-coupled to make a flip-flop: the output of each is wired as one of the inputs to the other.)

There is also no need to create the complement form of the sum u. However, the carry out of one bit position must be passed as the carry into the next bit position in both direct and complement forms. The most straightforward way to do this is to pass co through a 1-input NOR gate and label the output co2 , but that would add a gate delay in the worst possible place, slowing down the rippling of carries from right to left. An additional 4-input NOR gate building the canonical form of co2 (out of the opposite minterms as co) solves this problem.

The trade-off to maintain full speed in this way includes an unexpected cost (in addition to having to use a bigger gate). If we'd just used that 1-input gate to complement co, there would have been no use for the minterm m_7, and the gate that generated it could have been eliminated. Nevertheless, it's still a good trade.

Now we could have implemented those functions exactly according to their SoP and PoS canonical forms, by turning NOR gates into the functions specified. A NOR gate is made into an OR gate by passing its output through a 1-input NOR gate; and it is made into an AND gate by passing each of its inputs through a 1-input NOR gate. However, this approach not only increases the number of gates used, but also doubles the number of gate delays processing the signals, cutting the processing speed in half. Consequently, whenever performance is vital, going beyond canonical forms and doing the Boolean algebra to make the unenhanced NOR gates do the job is well worthwhile.

TOP-DOWN VS. BOTTOM-UP DESIGN

We have now seen how the minterm/maxterm tools can be used to design an adder stage in canonical form with the addition of some Boolean algebra, costing just 2 gate delays for each of the outputs. That's the "top-down" way to design the digital circuit for this function, but is it the best way? The discussion has focused on identifying "fastest" as "best," and the augmented canonical form meets that criterion flawlessly, but sometimes other factors predominate. The designer may have a primary goal of minimizing the number of gates, and/or of minimizing the fanouts of signals to other gates since big fanouts reduce resilience to a degraded power supply or other environmental factors. In such a case, a designer may develop the canonical-form design as a baseline, then try a bottom-up development, and finally compare the results.

The bottom-up development involves noticing that u = ci XOR (x XOR y), where XOR means eXclusive OR [true when either input is true but not when both are true], and that co = ci x + x y + y ci. One such development takes twelve NOR gates in all: six 2-input gates and two 1-input gates to produce u in 5 gate delays, plus three 2-input gates and one 3-input gate to produce co2 in 2 gate delays. The canonical baseline took eight 3-input NOR gates plus three 4-input NOR gates to produce u, co and co2 in 2 gate delays.

If the circuit inventory actually includes 4-input NOR gates, the top-down canonical design looks like a winner in both gate count and speed. But if (contrary to our convenient supposition) the circuits are actually 3-input NOR gates, of which two are required for each 4-input NOR function, then the canonical design takes 14 gates compared to 12 for the bottom-up approach, but still produces the sum digit u considerably faster. The fanout comparison is tabulated as:

Variables	Top-down	Bottom-up
x	4	1
x'	4	3
y	4	1
y'	4	3
ci	4	1
ci'	4	3
M or m	4@1,4@2	N/A
x XOR y	N/A	2
Misc	N/A	5@1
Max	4	3

The description of the bottom-up development mentions co2 as an output but not co. Does that design simply never need the direct form of the carry out? Well, yes and no. At each stage, the calculation of co2 depends only on ci2 , x2 and y2 , which means that the carry propagation ripples along the bit positions just as fast as in the canonical design without ever developing co.

The calculation of u, which does require ci to be made from ci2 by a 1-input NOR, is slower but for any word length the design only pays that penalty once (when the leftmost sum digit is developed). That's because those calculations overlap, each in what amounts to its own little pipeline without affecting when the next bit position's sum bit can be calculated. And, to be sure, the co2 out of the leftmost bit position will probably have to be complemented as part of the logic determining whether the addition overflowed. But using 3-input NOR gates, the bottom-up design is very nearly as fast for doing parallel addition on a non-trivial word length, cuts down on the gate count, and uses lower fanouts ... so it wins if gate count and/or fanout are paramount!

We'll leave the exact circuitry of the bottom-up design of which all these statements are true as an exercise for the interested reader, assisted by one more algebraic formula: u = ci(x XOR y) + ci2 (x XOR y)2]2 . Decoupling the carry propagation from the sum formation in this way is what elevates the performance of a carry-lookahead adder over that of a ripple carry adder.

Application in digital circuit design

One application of Boolean algebra is digital circuit design, with one goal to minimize the number of gates and another to minimize the settling time.

There are sixteen possible functions of two variables, but in digital logic hardware, the simplest gate circuits implement only four of them: conjunction (AND), disjunction (inclusive OR), and the respective complements of those (NAND and NOR).

Most gate circuits accept more than 2 input variables; for example, the spaceborne Apollo Guidance Computer, which pioneered the application of integrated circuits in the 1960s, was built with only one type of gate, a 3-input NOR, whose output is true only when all 3 inputs are false.

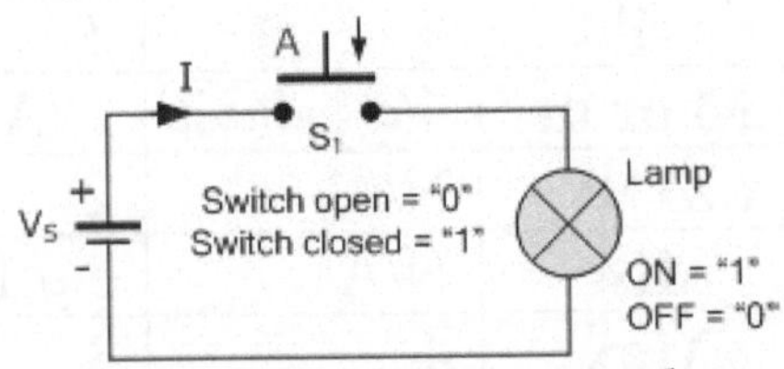

These logical states can be presented using electromechanical contacts in the form of switches or relays as a logic circuit element. The implementation of switching functions in digital logic circuits is nothing new, but it can give us a better understanding of how a single digital logic gate works.

Digital logic gates are the basic building blocks from which all digital electronic circuits and microprocessor based systems are made. They can be interconnected together to form either combinational logic circuits which are fully dependent on any external input signals applied to it, or sequential logic circuits which are dependent on its present stable state, feedback of its output, as well as any external input signals that may trigger a switching event.

The Switching Theory of a Switch

You may think that a switch is, well a switch, that can be used to turn a lighting load "ON" or "OFF". But a switch can also be a complex mechanical or electromechanical element used to control the flow of a signal through it in either direction, making it a bilateral device. Consider the circuit shown.

The Switching Theory of a Normally-open Switch

Here in this simple example, the lamp (L) is connected to the battery supply, V_S via the normally-open switch, S_1.

4

Number Systems and Conversion: Boolean Algebra

We have 10 fingers, so the number 10 comes naturally to us. We count and do math using decimal numbers. Another term for that is base-10. All that really means is that we use 10 unique digits and when we put those digits together to make bigger numbers, each place moving to the left is worth ten times the place to its right. We say there is the one's place, the ten's place, the hundred's place, the thousand's place, and so on.

The system concerned with the number and represented by a sequence of digits is called number system. It plays a vital role in computing and electronics. Number system also refers to the digits, its arrangement, positional value, and base of number system.

DIGITAL SYSTEMS AND SWITCHING CIRCUITS

Switching Theory is about using switches to implement Boolean expressions and logic gates for the the logic design of digital circuits

Switching Theory allows us to understand the operation and relationship between Boolean Algebra and two-level logic functions with regards to Digital Logic Gates. Switching theory can be used to further develop the theoretical knowledge and concepts of digital circuits when viewed as an interconnection of input elements producing an output state or condition.

Digital logic gates whose inputs and output can switch between two distinct logical values of 0 and 1, can be defined mathematically simply by using Boolean Algebra. But we can also represent the two digital logic states of HIGH or LOW, "1" or "0", "ON" or "OFF", as well as TRUE or FALSE.

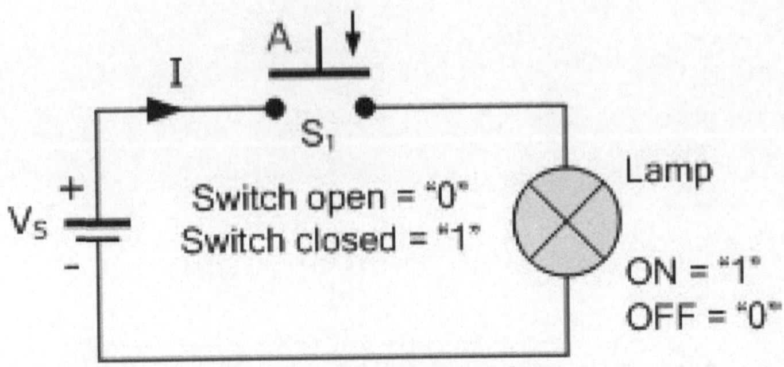

Thus if switch S_1 is not-pressed and therefore open, no current (I) flows so the lamp will be "OFF" and not illuminated. Likewise, if switch S_1 is pressed closing it, then current flows around the circuit and the lamp (L) will be "ON" and illuminated. Under normal steady state conditions the switch is permanently "open" so the lamp is "OFF".

We can use switching algebra to describe the operation of the circuit containing the switch, S_1. For example, if we label the normally-open switch as a variable with the letter "A", then when the switch is open, that is "A" is not-pressed, we can define the value of "A" as being "0". Again, when the switch is closed, that is "A" is pressed, we can define the value of "A" as being "1". This switching algebra theory is true for ALL normally-open switch configurations.

Switching Truth Table

We can develope this switching theory idea further by saying when the lamp is "ON" (illuminated), its switching alegebra variable will be "1", and when the lamp is "OFF" (not illuminated), its switching algebra variable will be "0".

A (Switch)	L (Lamp)
0	0
1	1
L = A	

Thus, when the switch is pressed (activated) the lamp is "ON", so "A" = 1 and "L" = 1, and when the switch is not-pressed (unactivated) the lamp is "OFF", so "A" = 0 and "L" = 0. Therefore we can correctly say that for the switching theory of the lamp, L = A as shown in the truth table.

The type of switch used in the above example is called a normally-open, make-contact switch as the have to physically make it for the switch to be considered closed (A = 1). But there is another type of switch arrangement which is the exact opposite in operation of the switch above called a normally-closed, break-contact switch which is constantly closed.

The Switching Theory of Series Switches

We have seen that the lamp (L) circuit above can be controlled using a single switch, S_1 and when S_1 is closed (pressed) current flows around the circuit and the lamp is "ON". But what if we added a second switch in series with S_1, how would that affect the switching function of the circuit and the illumination of the lamp.

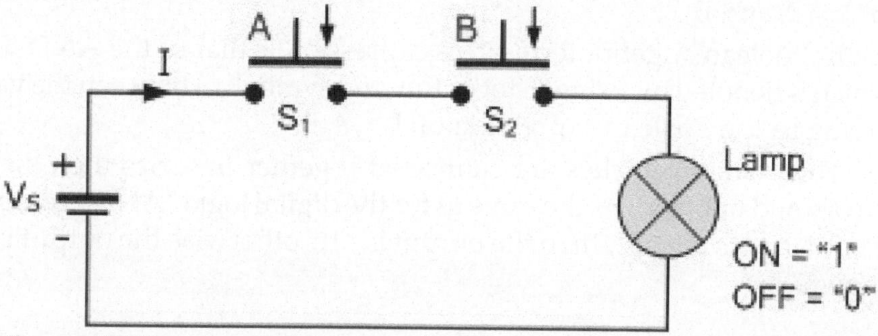

The switching circuit consists of two switches in series with a voltage source, V_S and the lamp. To distinquish the operation of each individual switch, we shall label switch, S_1 with the letter "A", and label switch, S_2 with the letter "B". Thus when either switch is open, that is not-pressed, we can define the value of "A" as being "0" and "B" as also being "0".

Likewise, when either switch is closed or pressed, we can define the value of "A" as being "1" or "B" as being "1". That is the logical level "1" corresponds to the supply voltage value, and will be positive. Whereas the logic level "0" corresponds to the voltage value of zero voltage, or ground.

As there are two switches, S_1 and S_2, or "A" and "B", then we can see that there are four possible combinations of the Boolean variables "A" and "B" to illuminate the lamp. For example, "A" is open and "B" is closed, or "A" is closed and "B" is open, or both "A" and "B" are open or closed at the same time. Then we can define these operations in the following switching theory truth table.

Series Switch Truth Table

The truth table shows that the lamp will only be "ON" and illuminated when BOTH switch, A AND switch, B are pressed and closed as pressing only one switch on its own will not cause current to flow.

A \ B	0	1
0	0	0
1	0	1
L = A and B		

This proves that when two switches S_1 and S_2 are connected in series, the only condition that will allow current (I) to flow and make the lamp illuminate is when both switches are closed giving the boolean expression of: L = A and B.

In Boolean Algebra terms, this expression is that of the AND function which is denoted by a single dot or full stop symbol, (.) between the variables giving us the Boolean expression of: L = A.B.

Thus when switches are connected together in series their switching theory and operation is the same as for the digital logic "AND" gate because if both inputs are "1", then the output is "1", otherwise the output is "0" as shown.

Digital Logic AND Gate

Thus if input "A" is AND'ed with input "B" it produces output "Q". In switching terms, the AND function is referred to as the Boolean Algebra multiplication function.

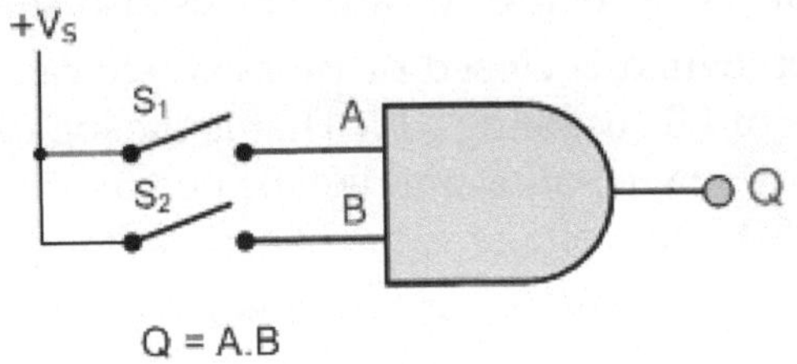

A	B	Q
0	0	0
0	1	0
1	0	0
1	1	1

The Switching Theory of Parallel Switches

If we now connect switches, S_1 and S_2 together in parallel as shown, how would is arrangement affect the switching function of the circuit and the illumination of the lamp.

The Switching Theory of Parallel Switches

The switching circuit now consists of the two switches in parallel with the voltage source, V_S and the lamp. As before, when either switch is open, that is not-pressed, we can define the value of "A" as being "0" and "B" as also being "0". Likewise, when either switch is closed or pressed, we can define the value of "A" as being "1" or "B" as being "1".

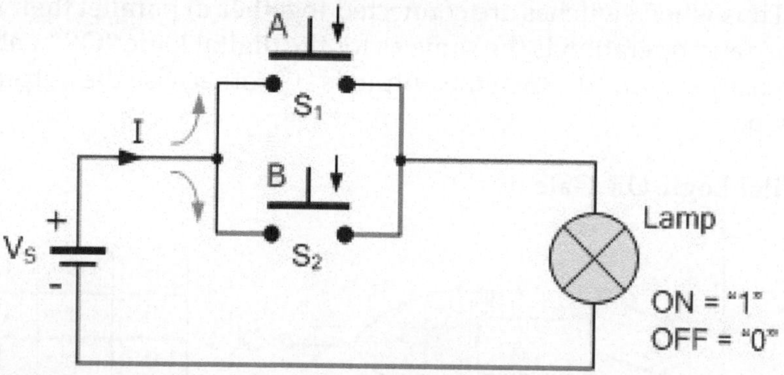

As before, with two switches, S_1 and S_2, or "A" and "B", there are four possible combinations of the Boolean variables "A" and "B" required to illuminate the lamp. The corresponding states are: "A" is open and "B" is closed, or "A" is closed and "B" is open, both "A" and "B" are open, or both closed at the same time. Then we can define these switching operations in the following switching theory truth table.

Parallel Switch Truth Table

The truth table shows that the lamp will only be "ON" and illuminated when EITHER switch, A OR switch, B are pressed and closed as pressing either switch will cause current to flow because there will always be a conducting path available for the lamp through whichever closed switch.

A \ B	0	1
0	0	1
1	1	1
L = A or B		

This therefore proves that when two switches S_1 and S_2 are connected together in parallel, the switching condition that allows current (I) to flow and make the lamp illuminate is when any one of the switches, or both are closed. This gives the boolean expression of: L = A or B.

In Boolean Algebra terms, this expression is that of the OR function which is denoted by a addition or plus sign, (+) between the variables giving us the Boolean expression of: L = A+B.

Thus when switches are connected together in parallel their switching theory and operation is the same as for the digital logic "OR" gate because if both inputs are "0", then the output is "0", otherwise the output is "1" as shown.

Digital Logic OR Gate

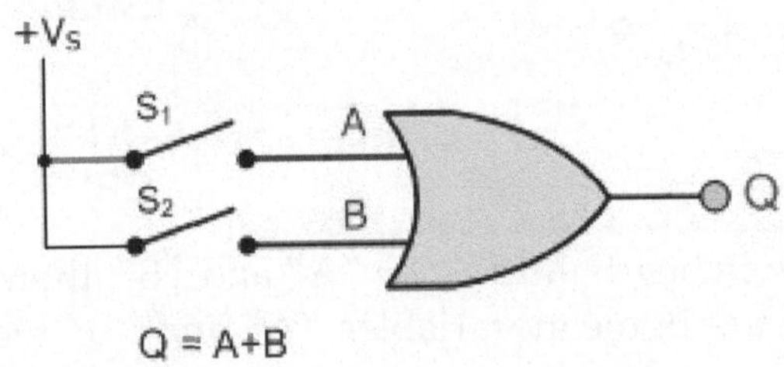

A	B	Q
0	0	0
0	1	1
1	0	1
1	1	1

Thus if input "A" is OR'ed with input "B" it produces output "Q", and in switching terms, the OR function is referred to as the Boolean Algebra logical addition function.

Switching Theory of a Boolean Function

Switching Theory can be used to implement Boolean expressions as well as digital logic gates. Ad we have seen above, in switch contact terms, the boolean expression using a dot (.) is interpreted as a series connection for Boolean multiplication, while a plus sign (+) is interpreted as a pair of parallel branches for Boolean addition.

Switching Theory Example No1

Implement the following Boolean function of Q = A(B+C) using switches to illuminate a lamp (or LED). Also show the equivalent digital logic circuit.

Switch Implementation

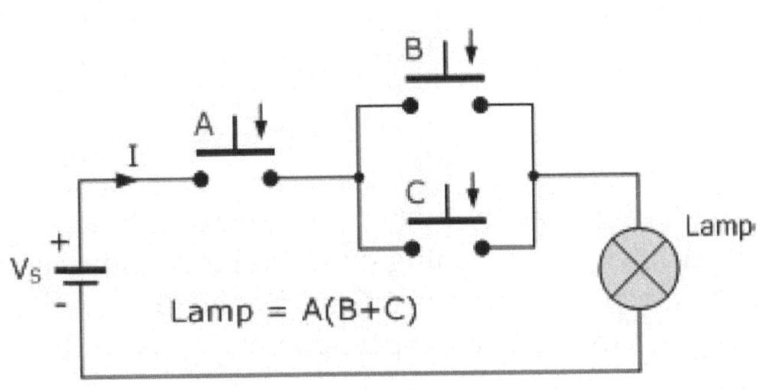

Logic Gate Implementation

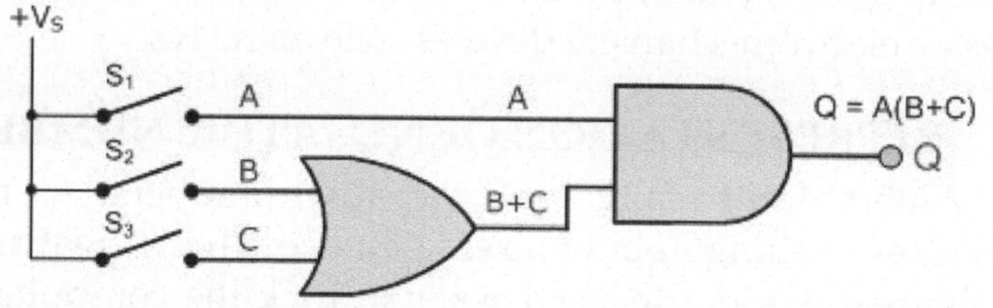

Indempotent Law of Switches

Thus far we have seen how to connect two switches together either in series or parallel to illuminate a lamp. But what if the two switches representing a Boolean AND function or an OR function (operations of multiplication and sum) are of the same single Boolean variable, A. In Boolean Algebra there are various laws and theroems which can be used to define the mathematics of logic circuits. One such theorem is known by the name of indempotent law.

Idempotent laws used in switching theory states that AND-ing or OR-ing a variable with itself will produce the original variable. For example variable "A" AND'ed with "A" will give "A", likewise variable "A" OR'ed with "A" will give "A", allowing us to simplify our switching circuits and we can demonstrate this below.

Indempotent Law of AND Function

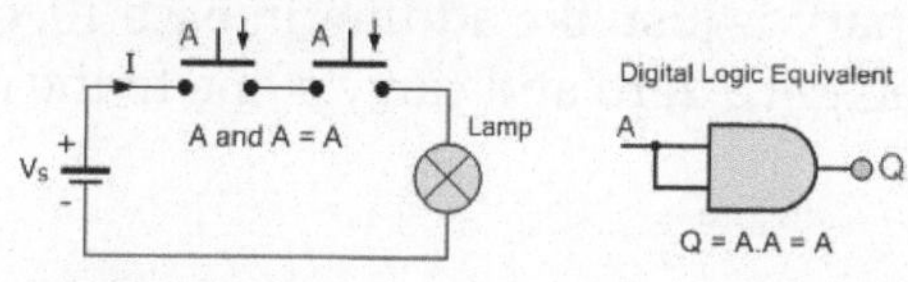

Indempotent Law of OR Function

We have seen here in this tutorial that switching theory techniques can be used to realise Boolean expressions and digital logic gate circuits using simple "ON/OFF" switches.

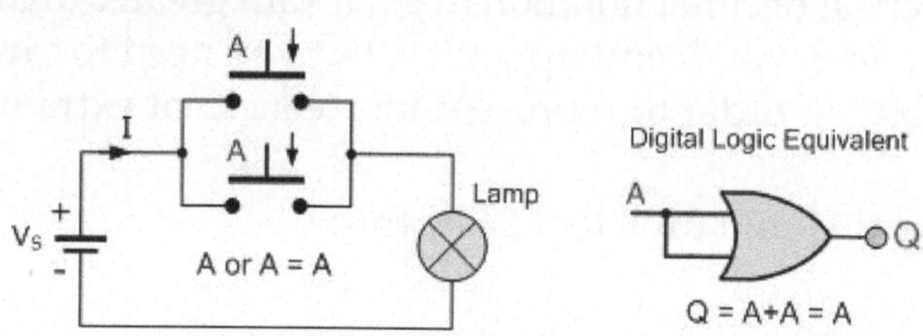

The representation of "AND" and "OR" functions using normally-open switches are easy to construct, easy to understand, and form the basic building blocks for most combinational logic circuits.

Thus given any Boolean expression or logic function, it is possible to use switching theory to implement it, after all, logic design is about using switches or electromechanical devices such as relays.

REPRESENTATION OF NEGATIVE NUMBERS

Computers don't really "get" negative numbers. As far as they're concerned, everything's either a zero or a one. That's great and all, but we live in the real world and need ways to trick the computer into storing negatives for us.

Signed Magnitude

If you just want to represent the number (without doing fancy math like adding and subtracting), include another 0 or 1 at the far left-hand side of the number as a "signed bit." The signed bit tells—you guessed it—the sign for the number: 0 for positive numbers and 1 for negative numbers. The catch? you can have a positive and negative version of zero.

Yeah, we don't get it either.

We left out a pretty important part of doing math on computers...actually doing any math. If you add -3 to 4, for example, you should get 1 as a result. Key word: should.

-3_{10} is represented as 1011 including the signed bit and 4^{10} is represented as 0100. After adding them together, you should get 0001.

Adding in binary is just like adding in base 10, except every time you add two ones, you get a zero and carry a one to the next place. Let's make a mini-algorithm:

- 0 + 0 = 0.
- 0 + 1 = 1.
- 1 + 1 = 0, carry the 1.

There's no way to represent two in one number in binary, just like there's no way to represent eleven with one number in base ten. It's just like adding two numbers in decimal notation to get a sum greater than ten. If you add together 6 and 8, you'll end up with 4, but you need to carry the extra 1 to the next place in order to represent fourteen. That extra number is called the carry.

Adding 1011 and 0100 looks like this:

$$\begin{array}{r} 1\ 0\ 1\ 1 \\ +\ 0\ 1\ 0\ 0 \\ \hline 1\ 1\ 1\ 1 \end{array} = -7_{10}$$

Good thing we have other systems to represent negative numbers in binary. Computers (and people) need to do math. Preferably correct math.

Two's Complement

To make binary numbers add together without, you know, ending up with -3 + 4 = -7 (can you hear the mathematicians cringe?), use two's complement.

Two's Complement could also be called "Signed Bits 2.0." In the opposite of signed bits, the leftmost value is reserved so that

- negative numbers start with a 0.
- positive numbers start with a 1.

This time, though, negative numbers are represented by flipping all of the bits—every single bit—in their positive counterparts. All the 0's will turn to 1's and all the 1's will turn to 0's. Then another 1 is added to the entire number.

Altogether, the process for manipulating -3 would be:

- Binary Three Plus the Signed Bit: 0011
- Bits Post-Flipping:: 1100
- Adding One: 1101

Now when you add -3 to 4, you get 10001.

	1				
1	0	1	0	0	
+	1	1	0	1	
1	0	0	0	1	$= 1_{10}$

Mission accomplished.

The extra one might look like it's going to make the number positive, but that isn't true. Instead, that extra number's going to be considered overflow, so it's just going to fall away from the four bit representation (everything inside the box).

This method works because you're essentially counting out how many numbers away from zero a negative number is—which is kind-of the whole point of negative numbers. For more on that, check out this page from Cornell.

The range of numbers you can write using this system goes from $-(2^{n-1})$ to $(2^{n-1} - 1)$ where n is the number of bits. In a four bit system (i.e. the one we used for the example) you can represent everything from $-(2^3)$ to $(2^3 - 1)$ or from -8_{10} to 7_{10}. An eight bit number system can range from -64 to 63. And…you get the idea.

Eight bits is also known as a byte. John Tukey, a mathematician known for his work in statistics, coined the term "bit" as an abbreviation of "binary digit" in the late 1940s.

EXPRESSIONS AND TRUTH TABLES

In designing digital circuits, the designer often begins with a truth table describing what the circuit should do.

The design task is largely to determine what type of circuit will perform the function described in the truth table.

While some people seem to have a natural ability to look at a truth table and immediately envision the necessary logic gate or relay logic circuitry for the task, there are procedural techniques available for the rest of us.

Here, Boolean algebra proves its utility in a most dramatic way. To illustrate this procedural method, we should begin with a realistic design problem. Suppose we were given the task of designing a flame detection circuit for a toxic waste incinerator. The intense heat of the fire is intended to neutralize the toxicity of the waste introduced into the incinerator.Such combustion-based techniques are commonly used to neutralize medical waste, which may be infected with deadly viruses or bacteria:

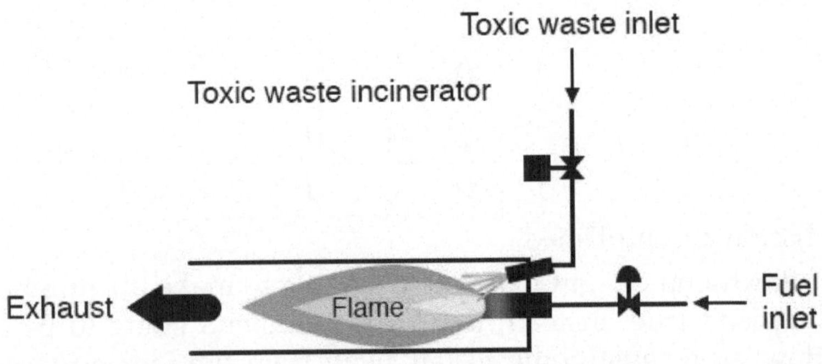

So long as a flame is maintained in the incinerator, it is safe to inject waste into it to be neutralized.

If the flame were to be extinguished, however, it would be unsafe to continue to inject waste into the combustion chamber, as it would exit the exhaust un-neutralized, and pose a health threat to anyone in close proximity to the exhaust.

What we need in this system is a sure way of detecting the presence of a flame, and permitting waste to be injected only if a flame is "proven" by the flame detection system.

Several different flame-detection technologies exist: optical (detection of light), thermal (detection of high temperature), and electrical conduction (detection of ionized particles in the flame path), each one with its unique advantages and disadvantages.

Suppose that due to the high degree of hazard involved with potentially passing un-neutralized waste out the exhaust of this incinerator, it is decided that the flame detection system be made redundant (multiple sensors), so that failure of a single sensor does not lead to an emission of toxins out the exhaust.

Each sensor comes equipped with a normally-open contact (open if no flame, closed if flame detected) which we will use to activate the inputs of a logic system:

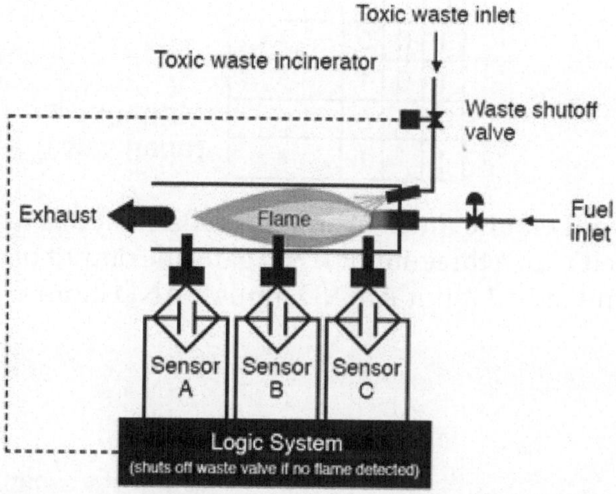

Our task, now, is to design the circuitry of the logic system to open the waste valve if and only if there is good flame proven by the sensors.

First, though, we must decide what the logical behavior of this control system should be.

Do we want the valve to be opened if only one out of the three sensors detects flame? Probably not, because this would defeat the purpose of having multiple sensors.

If any one of the sensors were to fail in such a way as to falsely indicate the presence of flame when there was none, a logic system based on the principle of "any one out of three sensors showing flame" would give the same output that a single-sensor system would with the same failure.

A far better solution would be to design the system so that the valve is commanded to open if and only if all three sensors detect a good flame.

This way, any single, failed sensor falsely showing flame could not keep the valve in the open position; rather, it would require all three sensors to be failed in the same manner—a highly improbable scenario—for this dangerous condition to occur.

Thus, our truth table would look like this:

Sensor Inputs (A, B, C)

A	B	C	Output
0	0	0	0
0	0	1	0
0	1	0	0
0	1	1	0
1	0	0	0
1	0	1	0
1	1	0	0
1	1	1	1

Output = 0 (close valve)

Output = 1 (open valve)

It does not require much insight to realize that this functionality could be generated with a three-input AND gate: the output of the circuit will be "high" if and only if input A AND input B AND input C are all "high:"

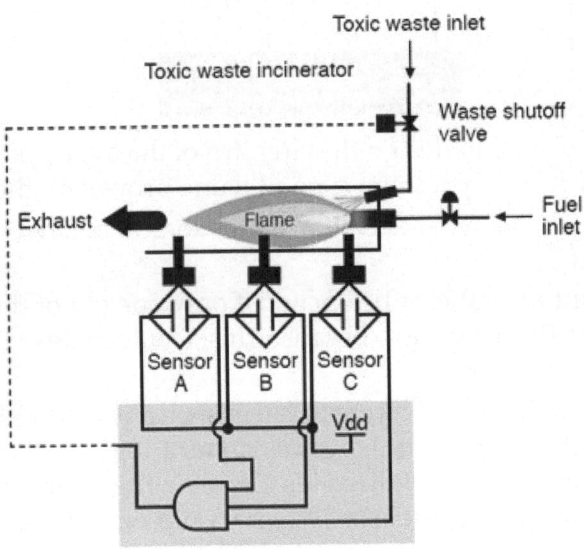

If using relay circuitry, we could create this AND function by wiring three relay contacts in series, or simply by wiring the three sensor contacts in series, so that the only way electrical power could be sent to open the waste valve is if all three sensors indicate flame:

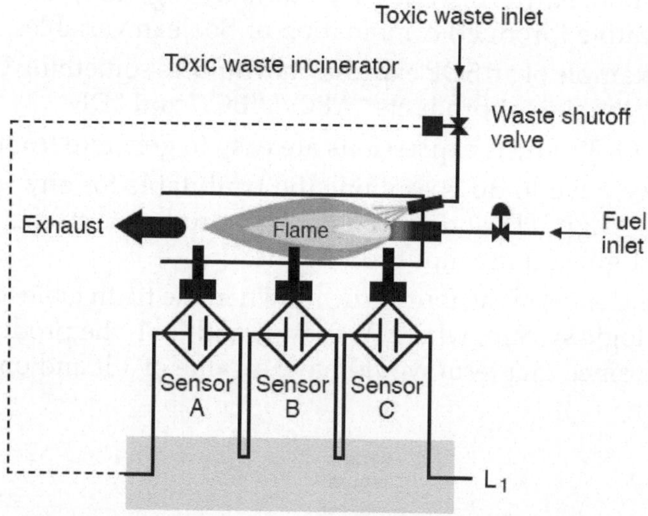

While this design strategy maximizes safety, it makes the system very susceptible to sensor failures of the opposite kind.

Suppose that one of the three sensors were to fail in such a way that it indicated no flame when there really was a good flame in the incinerator's combustion chamber.

That single failure would shut off the waste valve unnecessarily, resulting in lost production time and wasted fuel (feeding a fire that wasn't being used to incinerate waste).

It would be nice to have a logic system that allowed for this kind of failure without shutting the system down unnecessarily, yet still provide sensor redundancy so as to maintain safety in the event that any single sensor failed "high" (showing flame at all times, whether or not there was one to detect).

A strategy that would meet both needs would be a "two out of three" sensor logic, whereby the waste valve is opened if at least two out of the three sensors show good flame.

Using Sum-Of-Products

Here, it is not necessarily obvious what kind of logic circuit would satisfy the truth table.

However, a simple method for designing such a circuit is found in a standard form of Boolean expression called the Sum-Of-Products, or SOP, form.

As you might suspect, a Sum-Of-Products Boolean expression is literally a set of Boolean terms added (summed) together, each term being a multiplicative (product) combination of Boolean variables.

An example of an SOP expression would be something like this: ABC + BC + DF, the sum of products "ABC," "BC," and "DF."

Sum-Of-Products expressions are easy to generate from truth tables.

All we have to do is examine the truth table for any rows where the output is "high" (1), and write a Boolean product term that would equal a value of 1 given those input conditions.

For instance, in the fourth row down in the truth table for our two-out-of-three logic system, where A=0, B=1, and C=1, the product term would be A'BC, since that term would have a value of 1 if and only if A=0, B=1, and C=1:

Sensor Inputs

A	B	C	Output	
0	0	0	0	
0	0	1	0	
0	1	0	0	
0	1	1	1	$\overline{A}BC = 1$
1	0	0	0	
1	0	1	1	
1	1	0	1	
1	1	1	1	

Either one of these circuits will adequately perform the task of operating the incinerator waste valve based on a flame verification from two out of the three flame sensors.

At minimum, this is what we need to have a safe incinerator system.

We can, however, extend the functionality of the system by adding to it logic circuitry designed to detect if any one of the sensors does not agree with the other two.

If all three sensors are operating properly, they should detect flame with equal accuracy.

Thus, they should either all register "low" (000: no flame) or all register "high" (111: good flame).

Any other output combination (001, 010, 011, 100, 101, or 110) constitutes a disagreement between sensors, and may therefore serve as an indicator of a potential sensor failure.

If we added circuitry to detect any one of the six "sensor disagreement" conditions, we could use the output of that circuitry to activate an alarm.

Whoever is monitoring the incinerator would then exercise judgment in either continuing to operate with a possible failed sensor (inputs: 011, 101, or 110), or shut the incinerator down to be absolutely safe.

Also, if the incinerator is shut down (no flame), and one or more of the sensors still indicates flame (001, 010, 011, 100, 101, or 110) while the other(s) indicate(s) no flame, it will be known that a definite sensor problem exists.

The first step in designing this "sensor disagreement" detection circuit is to write a truth table describing its behavior.

Using Product-Of-Sums

An alternative to generating a Sum-Of-Products expression to account for all the "high" (1) output conditions in the truth table is to generate a Product-Of-Sums, or POS, expression, to account for all the "low" (0) output conditions instead.

Being that there are much fewer instances of a "low" output in the last truth table column, the resulting Product-Of-Sums expression should contain fewer terms.

As its name suggests, a Product-Of-Sums expression is a set of added terms (sums), which are multiplied (product) together.

An example of a POS expression would be (A + B)(C + D), the product of the sums "A + B" and "C + D".

To begin, we identify which rows in the last truth table column have "low" (0) outputs, and write a Boolean sum term that would equal 0 for that row's input conditions.

For instance, in the first row of the truth table, where A=0, B=0, and C=0, the sum term would be (A + B + C), since that term would have a value of 0 if and only if A=0, B=0, and C=0:

Good Flame: Output = 0 (close valve); Output = 1 (open valve)

Sensor Disagreement: Output = 0 (sensors agree); Output = 1 (sensors disagree)

Sensor Inputs A	B	C	Good Flame Output	Sensor Disagreement Output	
0	0	0	0	0	(A + B + C)
0	0	1	0	1	
0	1	0	0	1	
0	1	1	1	1	
1	0	0	0	1	
1	0	1	1	1	
1	1	0	1	1	
1	1	1	1	0	

Only one other row in the last truth table column has a "low" (0) output, so all we need is one more sum term to complete our Product-Of-Sums expression.

This last sum term represents a 0 output for an input condition of A=1, B=1 and C=1.

Therefore, the term must be written as (A′ + B′+ C′), because only the sum of the complemented input variables would equal 0 for that condition only:

Output = 0 (close valve) / Output = 1 (open valve) — Good Flame Output

Output = 0 (sensors agree) / Output = 1 (sensors disagree) — Sensor Disagreement Output

Sensor Inputs: A, B, C

A	B	C	Good Flame Output	Sensor Disagreement Output	
0	0	0	0	0	$(A + B + C)$
0	0	1	0	1	
0	1	0	0	1	
0	1	1	1	1	
1	0	0	0	1	
1	0	1	1	1	
1	1	0	1	1	
1	1	1	1	0	$(\bar{A} + \bar{B} + \bar{C})$

The completed Product-Of-Sums expression, of course, is the multiplicative combination of these two sum terms:

Output = 0 (close valve) / Output = 1 (open valve) — Good Flame Output

Output = 0 (sensors agree) / Output = 1 (sensors disagree) — Sensor Disagreement Output

Sensor Inputs: A, B, C

A	B	C	Good Flame Output	Sensor Disagreement Output	
0	0	0	0	0	$(A + B + C)$
0	0	1	0	1	
0	1	0	0	1	
0	1	1	1	1	
1	0	0	0	1	
1	0	1	1	1	
1	1	0	1	1	
1	1	1	1	0	$(\bar{A} + \bar{B} + \bar{C})$

$$\text{Output} = (A + B + C)(\bar{A} + \bar{B} + \bar{C})$$

Whereas a Sum-Of-Products expression could be implemented in the form of a set of AND gates with their outputs connecting to a single OR gate, a Product-Of-Sums expression can be implemented as a set of OR gates feeding into a single AND gate:

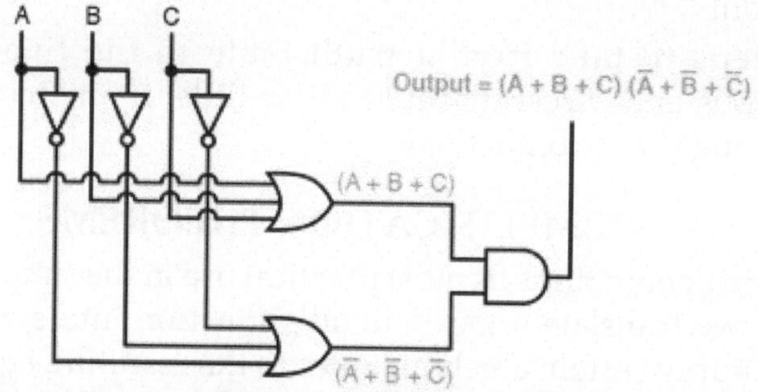

Correspondingly, whereas a Sum-Of-Products expression could be implemented as a parallel collection of series-connected relay contacts, a Product-Of-Sums expression can be implemented as a series collection of parallel-connected relay contacts:

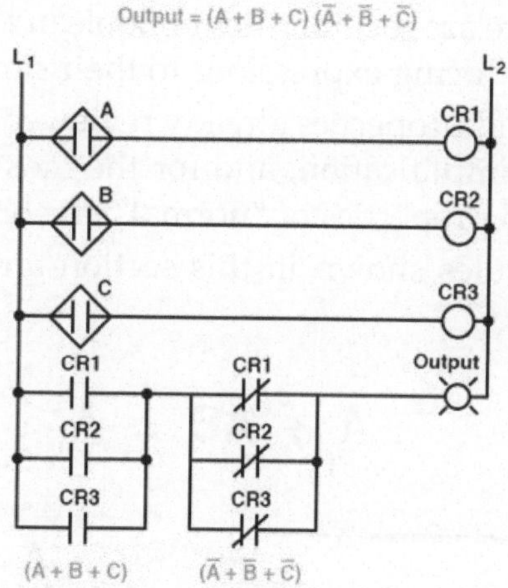

The previous two circuits represent different versions of the "sensor disagreement" logic circuit only, not the "good flame" detection circuit(s).

The entire logic system would be the combination of both "good flame" and "sensor disagreement" circuits, shown on the same diagram.

As you can see, both the Sum-Of-Products and Products-Of-Sums standard Boolean forms are powerful tools when applied to truth tables.

They allow us to derive a Boolean expression—and ultimately, an actual logic circuit—from nothing but a truth table, which is a written specification for what we want a logic circuit to do.

To be able to go from a written specification to an actual circuit using simple, deterministic procedures means that it is possible to automate the design process for a digital circuit. In other words, a computer could be programmed to design a custom logic circuit from a truth table specification!

The steps to take from a truth table to the final circuit are so unambiguous and direct that it requires little, if any, creativity or other original thought to execute them.

SIMPLIFICATION THEOREMS

Boolean algebra finds its most practical use in the simplification of logic circuits. If we translate a logic circuit's function into symbolic (Boolean) form, and apply certain algebraic rules to the resulting equation to reduce the number of terms and/or arithmetic operations, the simplified equation may be translated back into circuit form for a logic circuit performing the same function with fewer components.

If equivalent function may be achieved with fewer components, the result will be increased reliability and decreased cost of manufacture.

To this end, there are several rules of Boolean algebra presented in this section for use in reducing expressions to their simplest forms.

The identities and properties already reviewed in this chapter are very useful in Boolean simplification, and for the most part bear similarity to many identities and properties of "normal" algebra.

However, the rules shown in this section are all unique to Boolean mathematics.

$$A + AB = A$$

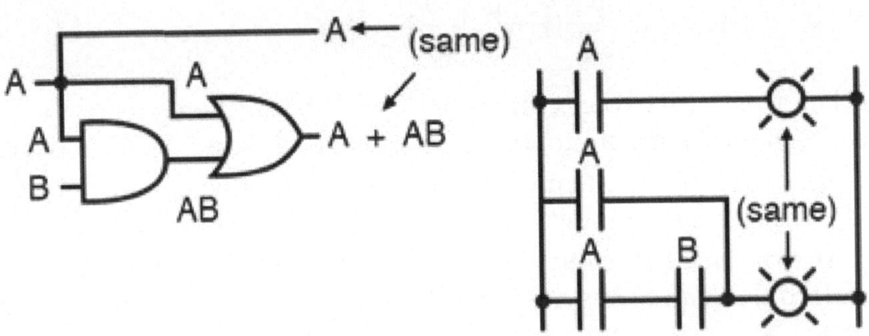

This rule may be proven symbolically by factoring an "A" out of the two terms, then applying the rules of A + 1 = 1 and 1A = A to achieve the final result:

$$A + AB$$

↓ Factoring A out of both terms

$$A(1 + B)$$

↓ Applying identity $A + 1 = 1$

$$A(1)$$

↓ Applying identity $1A = A$

$$A$$

Please note how the rule A + 1 = 1 was used to reduce the (B + 1) term to 1.

When a rule like "A + 1 = 1" is expressed using the letter "A", it doesn't mean it only applies to expressions containing "A".

What the "A" stands for in a rule like A + 1 = 1 is any Boolean variable or collection of variables.

This is perhaps the most difficult concept for new students to master in Boolean simplification: applying standardized identities, properties, and rules to expressions not in standard form.

For instance, the Boolean expression ABC + 1 also reduces to 1 by means of the "A + 1 = 1" identity.

In this case, we recognize that the "A" term in the identity's standard form can represent the entire "ABC" term in the original expression.

The next rule looks similar to the first one shown in this section, but is actually quite different and requires a more clever proof:

$$A + \bar{A}B = A + B$$

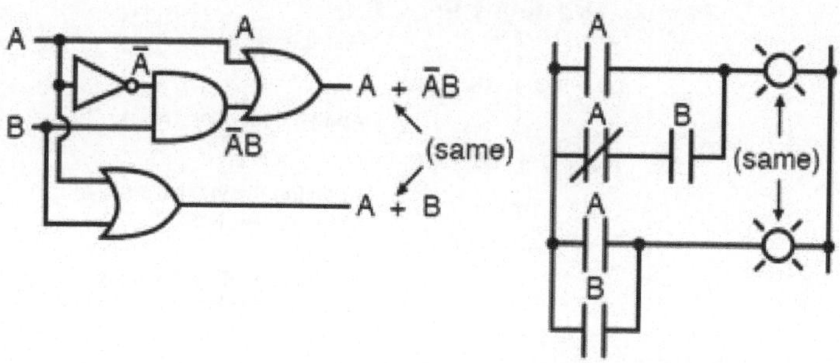

$A + \bar{A}B$

↓ Applying the previous rule to expand A term $A + AB = A$

$A + AB + \bar{A}B$

↓ Factoring B out of 2nd and 3rd terms

$A + B(A + \bar{A})$

↓ Applying identity $A + \bar{A} = 1$

$A + B(1)$

↓ Applying identity $1A = A$

$A + B$

Note how the last rule (A + AB = A) is used to "un-simplify" the first "A" term in the expression, changing the "A" into an "A + AB".

While this may seem like a backward step, it certainly helped to reduce the expression to something simpler!

Sometimes in mathematics we must take "backwar≤ steps to achieve the most elegant solution.

Knowing when to take such a step and when not to is part of the art-form of algebra, just as a victory in a game of chess almost always requires calculated sacrifices.

Another rule involves the simplification of a product-of-sums expression:

$$(A + B)(A + C) = A + BC$$

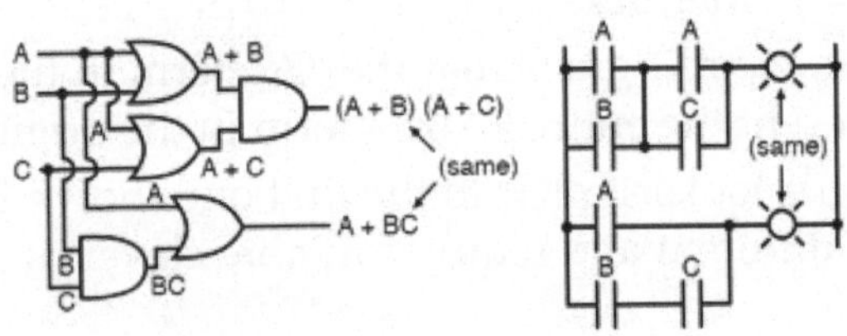

And, the corresponding proof:

$(A + B)(A + C)$

↓ Distributing terms

$AA + AC + AB + BC$

↓ Applying identity $AA = A$

$A + AC + AB + BC$

↓ Applying identity $A + AB = A$ to the $A + AC$ term

$A + AB + BC$

↓ Applying identity $A + AB = A$ to the $A + AB$ term

$A + BC$

To summarize, here are the three new rules of Boolean simplification expounded in this section:

Useful Boolean rules for simplification

$$A + AB = A$$

$$A + \bar{A}B = A + B$$

$$(A + B)(A + C) = A + BC$$

Simple Rules of 0 and 1 (best name for them I guess):

- Annihilator of Multiplication: Anything that multiplies (AND) with 0 gives 0 (A * 0 = 0)
- Annihilator of Addition: Anything that adds (OR) with 1 gives 1 (A + 1 = 1)
- Identity for Multiplication: Anything that multiplies (AND) with 1 gives itself (A * 1 = A)
- Identity for Addition: Anything that adds (OR) with 0 gives itself (A + 0 = A)

All except the second one hold in Ordinary and Boolean Algebra.

Boolean Algebra only Laws

- Idempotence: Add or Multiply with self gives self (AA = A and A + A = A)
- Absorption: Special Case of Distributivity [A(A + B) = A and A + AB = A]
- Distributivity of Addition over Multiplication [A + BC = AB + AC]

Complement Laws

- Anything multiplied (AND) with its complement gives 0 (A*A' = 0)
- Anything added (AND) with its complement gives 1(A + A' = 1)
- If we complement two times we get ourself again (A'' = A)
- Using the Truth Table we know that A + A'B = A + B and A' + AB = A' + B. So, when having both the "one" that is alone survives and the "one" that is together with another variable gets cleared!

De Morgan Theorem

The De Morgan Theorem is used to make our Circuit even smaller including more gates then the AND, OR, NOT Gates. Using the Universal NOR and NAND Gates we can then simplify our Circuit even further (We will talk about them separatelly some posts later on)

The Theorem for two variables looks like this:

- (AB)′ = A′ + B′
- (A + B)′ = A′B′

A simple way of memorizing it is that AND turns into OR and vise versa, and A turns into NOT A or A′ and vise versa.

So, we can generalize it for even more:

- (ABC)′ = A′ + B′ + C′
- (A + B + C)′ = A′B′C′

Using this Theorem we can get rid of all the Complements putting them all together and making them into an NAND or NOR Gate [NAND is actually (AB)′ and NOR is actually (A+B)′].

Example Simplification

Suppose we have a circuit and we want to simplify it. The circuit function of that logical circuit contains only basic Gates (AND, OR, NOT) and we want to keep using only those basic Gates.

The function looks like this:

F(A, B, C) = (A+B)B′ + B′ + BC

We use the Distributivity Law [(A+B)B′ = AB′ + BB′]:

F(A, B, C) = AB′ + BB′ + B′ + BC

We use one of the Complement Law(BB′ = 0) and afterwards the Identity of Addition (AB′ + 0 = AB′):

F(A, B, C) = AB′ + B′ + BC

We take a common factor B′ of the first two:

F(A, B, C) = B′(A + 1) + BC

We use the Annihilator of Addition (A + 1 = 1) and afterwards the Identity of Multiplication (B ′* 1 = B′):

F(A, B, C) = B′ + BC

We use the Complement Law (B′ + BC = B′ + C):

F(A, B, C) = B′ + C

The last expression is the simplified function. As you can see we don't have A as an input anymore cause it is not needed for getting the specific output values that this circuit gave us!

So, our Circuit now contains only two Gates: An OR Gate that connects the two Inputs and an NOT Gate that inverts the B Input's value!

Boolean Algebraic Properties

Another type of mathematical identity, called a "property" or a "law," describes how differing variables relate to each other in a system of numbers.

The Commutative Property

One of these properties is known as the commutative property, and it applies equally to addition and multiplication. In essence, the commutative property tells us we can reverse the order of variables that are either added together or multiplied together without changing the truth of the expression:

Commutative Property of Addition

$$A + B = B + A$$

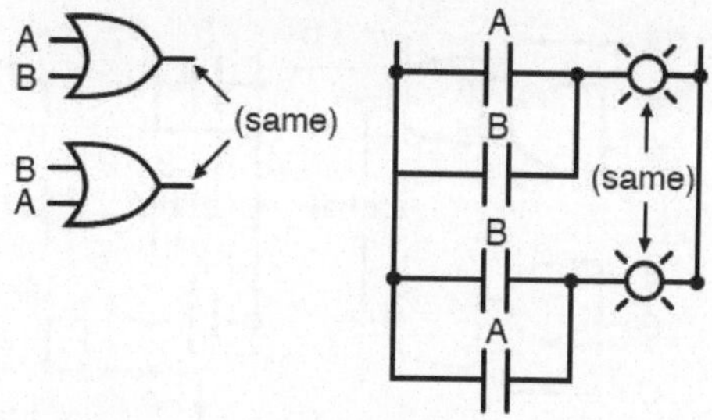

Commutative Property of Multiplication

$$AB = BA$$

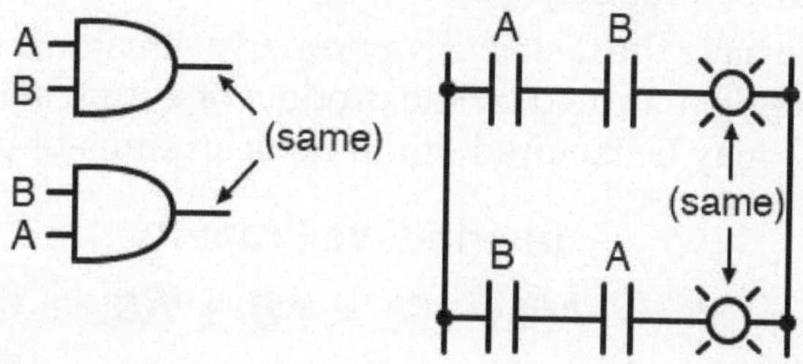

The Associative Property

Along with the commutative properties of addition and multiplication, we have the associative property, again applying equally well to addition and multiplication.

This property tells us we can associate groups of added or multiplied variables together with parentheses without altering the truth of the equations.

Associative Property of Addition

$$A + (B + C) = (A + B) + C$$

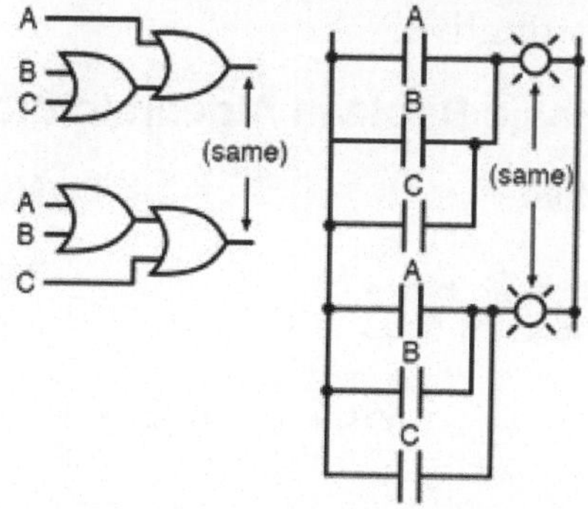

Associative Property of Multiplication

$A(BC) = (AB)C$

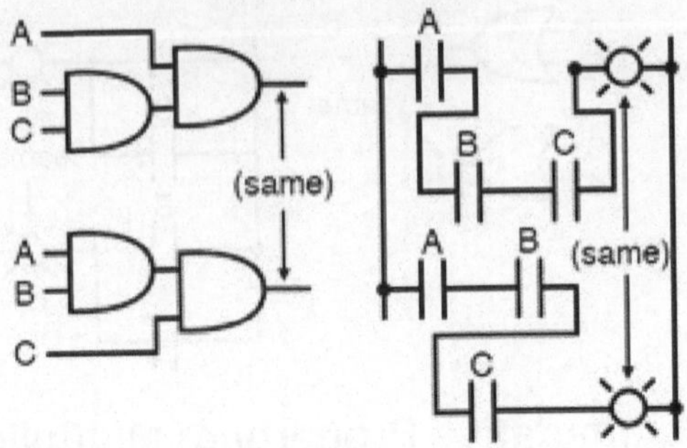

The Distributive Property

Lastly, we have the distributive property, illustrating how to expand a Boolean expression formed by the product of a sum, and in reverse shows us how terms may be factored out of Boolean sums-of-products:

Distributive Property

$A(B + C) = AB + AC$

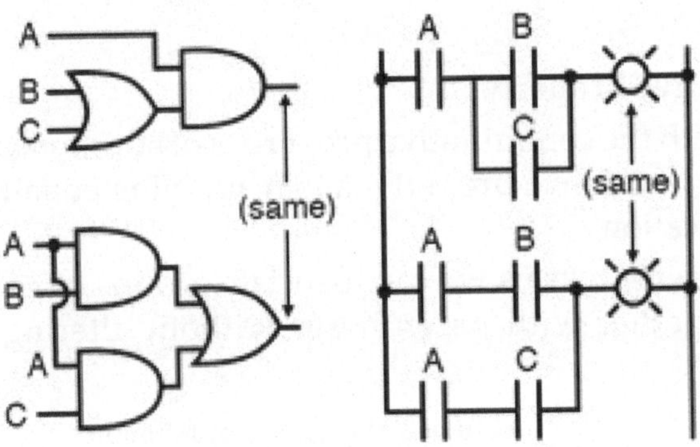

To summarize, here are the three basic properties: commutative, associative, and distributive.

Basic Boolean Algebraic Properties

Additive	Multiplicative
$A + B = B + A$	$AB = BA$
$A + (B + C) = (A + B) + C$	$A(BC) = (AB)C$

$$A(B + C) = AB + AC$$

Examples

- Every finite Boolean algebra is complete.
- The algebra of subsets of a given set is a complete Boolean algebra.
- The regular open sets of any topological space form a complete Boolean algebra. This example is of particular importance because every forcing poset can be considered as a topological space (a base for the topology consisting of sets that are the set of all elements less than or equal to a given element). The corresponding regular open algebra can be used to form Boolean-valued models which are then equivalent to generic extensions by the given forcing poset.
- The algebra of all measurable subsets of a ó-finite measure space, modulo null sets, is a complete Boolean algebra. When the measure space is the unit interval with the ó-algebra of Lebesgue measurable sets, the Boolean algebra is called the random algebra.
- The algebra of all measurable subsets of a measure space is a 5!$_1$-complete Boolean algebra, but is not usually complete.
- The algebra of all subsets of an infinite set that are finite or have finite complement is a Boolean algebra but is not complete.
- The Boolean algebra of all Baire sets modulo meager sets in a topological space with a countable base is complete; when the topological space is the real numbers the algebra is sometimes called the Cantor algebra.
- Another example of a Boolean algebra that is not complete is the Boolean algebra P(ù) of all sets of natural numbers, quotiented out by the ideal Fin of finite subsets. The resulting object, denoted P(ù)/Fin, consists of all equivalence classes of sets of naturals, where the relevant equivalence relation is that two sets of naturals are equivalent if their symmetric difference is finite. The Boolean operations are defined analogously, for example, if A and B are two equivalence classes in P(ù)/Fin, we define A '" B to be the equivalence class of a)" b, where a and b are some (any) elements of A and B respectively.

Now let a_0, a_1, ... be pairwise disjoint infinite sets of naturals, and let A_0, A_1, ... be their corresponding equivalence classes in P(ù)/Fin. Then given any upper bound X of A_0, A_1, ... in P(ù)/Fin, we can find a lesser upper bound, by removing from a representative for X one element of each a_n. Therefore the A_n have no supremum.

- A Boolean algebra is complete if and only if its Stone space of prime ideals is extremally disconnected.

Properties of complete Boolean algebras

- Sikorski's extension theorem states that if A is a subalgebra of a Boolean algebra B, then any homomorphism from A to a complete Boolean algebra C can be extended to a morphism from B to C.

- Every subset of a complete Boolean algebra has a supremum, by definition; it follows that every subset also has an infimum (greatest lower bound).
- For a complete boolean algebra both infinite distributive laws hold.
- For a complete boolean algebra infinite de-Morgan's laws hold.

THE COMPLETION OF A BOOLEAN ALGEBRA

The completion of a Boolean algebra can be defined in several equivalent ways:

- The completion of A is (up to isomorphism) the unique complete Boolean algebra B containing A such that A is dense in B; this means that for every nonzero element of B there is a smaller non-zero element of A.
- The completion of A is (up to isomorphism) the unique complete Boolean algebra B containing A such that every element of B is the supremum of some subset of A.

The completion of a Boolean algebra A can be constructed in several ways:

- The completion is the Boolean algebra of regular open sets in the Stone space of prime ideals of A. Each element x of A corresponds to the open set of prime ideals not containing x (which is open and closed, and therefore regular).
- The completion is the Boolean algebra of regular cuts of A. Here a cut is a subset U of A^+ (the non-zero elements of A) such that if q is in U and $p \leq q$ then p is in U, and is called regular if whenever p is not in U there is some $r \leq p$ such that U has no elements $\leq r$. Each element p of A corresponds to the cut of elements $\leq p$.

If A is a metric space and B its completion then any isometry from A to a complete metric space C can be extended to a unique isometry from B to C. The analogous statement for complete Boolean algebras is not true: a homomorphism from a Boolean algebra A to a complete Boolean algebra C cannot necessarily be extended to a (supremum preserving) homomorphism of complete Boolean algebras from the completion B of A to C. (By Sikorski's extension theorem it can be extended to a homomorphism of Boolean algebras from B to C, but this will not in general be a homomorphism of complete Boolean algebras; in other words, it need not preserve suprema.)

Free ê-complete Boolean algebras

Unless the Axiom of Choice is relaxed, free complete boolean algebras generated by a set do not exist (unless the set is finite). More precisely, for any cardinal ê, there is a complete Boolean algebra of cardinality $2^{ê}$ greater than ê that is generated as a complete Boolean algebra by a countable subset; for example the Boolean algebra of regular open sets in the product space ê^{ù}, where ê has the discrete topology. A countable generating set consists of all sets $a_{m,n}$ for m, n integers, consisting of the elements x ″ ê^{ù} such that x(m) < x(n). (This boolean algebra is called a collapsing algebra, because forcing with it collapses the cardinal ê onto ù.)

In particular the forgetful functor from complete Boolean algebras to sets has no left adjoint, even though it is continuous and the category of Boolean algebras is small-complete. This shows that the "solution set condition" in Freyd's adjoint functor theorem is necessary.

Given a set X, one can form the free Boolean algebra A generated by this set and then take its completion B. However B is not a "free" complete Boolean algebra generated by X (unless X is finite or AC is omitted), because a function from X to a free Boolean algebra C cannot in general be extended to a (supremum-preserving) morphism of Boolean algebras from B to C. On the other hand, for any fixed cardinal ê, there is a free (or universal) ê-complete Boolean algebra generated by any given set.

5

Topics in the Theory of Boolean Algebra

The axioms of a ring give structure to the operations of addition and multiplication on a set. However, we can construct algebraic structures, known as lattices and Boolean algebras, that generalize other types of operations. For example, the important operations on sets are inclusion, union, and intersection. Lattices are generalizations of order relations on algebraic spaces, such as set inclusion in set theory and inequality in the familiar number systems $\mathbb{N}, \mathbb{Z}, \mathbb{Q}$, and $\mathbb{R}$. Boolean algebras generalize the operations of intersection and union. Lattices and Boolean algebras have found applications in logic, circuit theory, and probability.

LATTICES

Partially Ordered Sets

We begin the study of lattices and Boolean algebras by generalizing the idea of inequality. Recall that a relation on a set X is a subset of X × X. A relation on X is called a partial order of X if it satisfies the following axioms.

1. The relation is reflexive: (a,a)″P for all a″X.
2. The relation is antisymmetric: if (a,b)″P and (b,a)″P, then a=b.
3. The relation is transitive: if (a,b)″P and (b,c)″P, then (a,c)″P.

We will usually write to mean unless some symbol is naturally associated with a particular partial order, such as with integers and , or with sets and . A set together with a partial order is called a partially ordered set, or poset.

Example 19.1

The set of integers (or rationals or reals) is a poset where

Solution

a≤b has the usual meaning for two integers a and b in Z.

ATOMS

Given an element a of a Boolean algebra (or other poset) A, recall that a is an atom in A if a is minimal among non-trivial (non-bottom) elements of A. That is, given any b″A in A such that b≤a, either b=0 or b=a.

A Boolean algebra A is atomic if we have b=Á″$_I$ a_i for every b″A, where $\{a_i\}$ is some set of atoms in A.

Boolean algebras not isomorphic to power sets

An atom of a Boolean algebra is an element x such that there exist exactly two elements y satisfying y ≤ x, namely x and 0. A Boolean algebra is said to be atomic when every element is a sup of some set of atoms (the bottom element is always the empty sup).

Every finite Boolean algebra is atomic, and moreover isomorphic to the power set 2^X of the set X of its atoms, under the operations of union, intersection, and complement, with 0 and 1 realized by respectively the empty set and X. Conversely every finite power set forms a Boolean algebra under union, intersection, and complement.

This converse can be strengthened by dropping "finite." However the converse of this strengthening does not hold: there exist Boolean algebras that are not isomorphic to any power set under union and intersection. The smallest infinite power set is 2^N where N is the set of natural numbers, which Cantor showed to be uncountable. There do however exist countably infinite Boolean algebras, two examples of which we now give.

1. Finite and confinite sets of integers. A cofinite set of integers is one that omits only finitely many integers. The finite sets are clearly closed under finite union and finite intersection, and likewise the cofinite sets are so closed. The complement (with respect to the set of all integers) of a finite set is cofinite and vice versa. The union of a finite set with a cofinite set is cofinite, while their intersection is finite. Hence this set of sets of integers is closed under all the Boolean operations. These operations furthermore satisfy all the Boolean identities. Hence this set of sets forms a field of sets in the above sense of Birkhoff and hence a Boolean algebra. It is atomic (the atoms are the singletons or one-element sets, and every set is the sup of its elements in this algebra) but not complete (any set X of atoms has a sup if and only if the union of X is either finite or cofinite, so in particular the set of even atoms has no sup).

2. The free Boolean algebra on countably many generators. Given countably many variables $v_0, v_1, v_2, \ldots$, this is the set of all abstract terms that

can be built up from these variables using the operations x∧y, x("y, ¬x, and the constants 0 and 1. Any such term will be finite and so use only finitely many variables. Terms are abstract in the sense that two terms that always have the same value in 2 are considered to be the same term, e.g. x∧y and y∧x. Without the abstractness requirement, x∧y and y∧x would be distinct terms whence they would violate the Boolean identity x∧y = y∧x when built up from x and y. The abstractness condition ensures that the Boolean identities are always satisfied.

The free countable Boolean algebra is atomless. In particular, for any x there is always a smaller nonzero element, namely $x \wedge v_i$ where v_i is any variable not appearing in the term x. Surprisingly it can be shown that every countable atomless Boolean algebra is isomorphic to the free countable Boolean algebra. Furthermore this algebra is not complete, for example any set X of variables has a sup if and only if the set is finite.

The free Boolean algebra on n generators, n finite, has 2^{2n} elements. These are the semantically distinct terms that can be built using n variables. When n = 2 this algebra has 16 elements, namely the 16 binary Boolean operations. Because finitely generated Boolean algebras are finite, the class of Boolean algebras is said to be locally finite. (Contrast this with finitely generated monoids, e.g. the set of all words on a 2-letter alphabet; the class of monoids is therefore not locally finite.)

SYMMETRIC DIFFERENCE: BOOLEAN RINGS

An associative ring K whose elements are all idempotent, i.e. $x^2 = x$ for any x " K. Any Boolean ring K `" 0 is commutative and is a subdirect sum of fields $Z^2$ of two elements, and x + x = 0 for all x " K. A finite Boolean ring K `" 0 is a direct sum of fields Z^2 and therefore has a unit element.

A Boolean ring is the ring version of a Boolean algebra, namely: Any Boolean algebra is a Boolean ring with a unit element under the operations of addition and multiplication defined by the rules

(x+y) = (x)"Cy)*"(y)"Cx), x.y = x)"y,

where Cx is the complement of x. The zero and the unit of the ring are the same as, respectively, the zero and the unit of the algebra. Conversely, every Boolean ring with a unit element is a Boolean algebra under the operations x*"y = x+y+xy, x)"y = x.y, Cx = 1+x.

Notations

There are at least four different and incompatible systems of notation for Boolean rings and algebras:

- In commutative algebra the standard notation is to use x + y = (x ∧ ¬ y) (" (¬ x ∧ y) for the ring sum of x and y, and use xy = x ∧ y for their product.

- In logic, a common notation is to use x ∧ y for the meet (same as the ring product) and use x (“ y for the join, given in terms of ring notation (given just above) by x + y + xy.
- In set theory and logic it is also common to use x · y for the meet, and x + y for the join x (“ y. This use of + is different from the use in ring theory.
- A rare convention is to use xy for the product and x ⊕ y for the ring sum, in an effort to avoid the ambiguity of +.

Historically, the term "Boolean ring" has been used to mean a "Boolean ring possibly without an identity", and "Boolean algebra" has been used to mean a Boolean ring with an identity. The existence of the identity is necessary to consider the ring as an algebra over the field of two elements: otherwise there cannot be a (unital) ring homomorphism of the field of two elements into the Boolean ring. (This is the same as the old use of the terms "ring" and "algebra" in measure theory.[a])

EXAMPLES

One example of a Boolean ring is the power set of any set X, where the addition in the ring is symmetric difference, and the multiplication is intersection. As another example, we can also consider the set of all finite or cofinite subsets of X, again with symmetric difference and intersection as operations. More generally with these operations any field of sets is a Boolean ring. By Stone's representation theorem every Boolean ring is isomorphic to a field of sets (treated as a ring with these operations).

Relation to Boolean algebras

Since the join operation (“ in a Boolean algebra is often written additively, it makes sense in this context to denote ring addition by ⊕, a symbol that is often used to denote exclusive or.

Given a Boolean ring R, for x and y in R we can define

x ‘“ y = xy,

x (“ y = x ⊕ y ⊕ xy,

¬x = 1 ⊕ x.

These operations then satisfy all of the axioms for meets, joins, and complements in a Boolean algebra. Thus every Boolean ring becomes a Boolean algebra. Similarly, every Boolean algebra becomes a Boolean ring thus:

xy = x ‘“ y,

x ⊕ y = (x (“ y) ‘“ ¬(x ‘“ y).

If a Boolean ring is translated into a Boolean algebra in this way, and then the Boolean algebra is translated into a ring, the result is the original ring. The analogous result holds beginning with a Boolean algebra. A map between two Boolean rings is a ring homomorphism if and only if it is a homomorphism of the corresponding Boolean algebras. Furthermore, a subset of a Boolean ring is a ring ideal (prime ring ideal, maximal ring ideal) if and only if it is an order ideal (prime order ideal, maximal order ideal) of the Boolean algebra. The quotient ring of a Boolean ring modulo a ring ideal corresponds to the factor algebra of the corresponding Boolean algebra modulo the corresponding order ideal.

PROPERTIES OF BOOLEAN RINGS

Every Boolean ring R satisfies $x \oplus x = 0$ for all x in R, because we know

$$x \oplus x = (x \oplus x)^2 = x^2 \oplus x^2 \oplus x^2 \oplus x^2 = x \oplus x \oplus x \oplus x$$

and since $(R,\oplus)$ is an abelian group, we can subtract $x \oplus x$ from both sides of this equation, which gives $x \oplus x = 0$. A similar proof shows that every Boolean ring is commutative:

$$x \oplus y = (x \oplus y)^2 = x^2 \oplus xy \oplus yx \oplus y^2 = x \oplus xy \oplus yx \oplus y$$

and this yields $xy \oplus yx = 0$, which means $xy = yx$ (using the first property above).

The property $x \oplus x = 0$ shows that any Boolean ring is an associative algebra over the field F_2 with two elements, in precisely one way. In particular, any finite Boolean ring has as cardinality a power of two. Not every unital associative algebra over F_2 is a Boolean ring: consider for instance the polynomial ring $F_2[X]$.

The quotient ring R/I of any Boolean ring R modulo any ideal I is again a Boolean ring. Likewise, any subring of a Boolean ring is a Boolean ring.

Any localization RS^{-1} of a Boolean ring R by a set $S \subseteq R$ is a Boolean ring, since every element in the localization is idempotent.

The maximal ring of quotients Q (R) (in the sense of Utumi and Lambek) of a Boolean ring R is a Boolean ring, since every partial endomorphism is idempotent.

Every prime ideal P in a Boolean ring R is maximal: the quotient ring R/P is an integral domain and also a Boolean ring, so it is isomorphic to the field F_2, which shows the maximality of P. Since maximal ideals are always prime, prime ideals and maximal ideals coincide in Boolean rings.

Every finitely generated ideal of a Boolean ring is principal (indeed, $(x,y) = (x + y + xy)$). Furthermore, as all elements are idempotents, Boolean rings are commutative von Neumann regular rings and hence absolutely flat, which means that every module over them is flat.

Unification

Unification in Boolean rings is decidable, that is, algorithms exist to solve arbitrary equations over Boolean rings. Both unification and matching in finitely generated free Boolean rings are NP-complete, and both are NP-hard in finitely presented Boolean rings. (In fact, as any unification problem f(X) = g(X) in a Boolean ring can be rewritten as the matching problem f(X) + g(X) = 0, the problems are equivalent.)

Unification in Boolean rings is unitary if all the uninterpreted function symbols are nullary and finitary otherwise (i.e. if the function symbols not occurring in the signature of Boolean rings are all constants then there exists a most general unifier, and otherwise the minimal complete set of unifiers is finite).

Ideals

Let A be a Boolean algebra and B a subset of A. The following are equivalent:

1. If A is interpreted as a Boolean ring, B is a ring ideal.
2. If A is interpreted as a Boolean lattice, B is a lattice ideal.

Before proving this equivalence, we want to mention that a Boolean ring is equivalent to a Boolean lattice, and a more commonly used terminology is a Boolean algebra, which is also valid, as it is an algebra over the ring of integers. The standard way of characterizing the ring structure from the lattice structure is by defining a+b:=(a2 '"b)("(a∧b2) (called the symmetric difference) and a.b = a∧b. From this, we can "solve" for (" in terms of + and .: a("b=a+b+a.b.

Proof. First, suppose B is an ideal of the "ring" A. If a,b"B, then a("b=a+b+a.b"B. Suppose now that a"B and c " A with c ≤ a. Then c = c∧a = c.a " B as well. So B is a lattice ideal of A.

Next, suppose B is an ideal of the "lattice" A. If a,b"B, then both a2 '"b and a∧b2 are in B since the first one is less than or equal to b and the second less than or equal to a, so their join is in B as well, this means that a"b=a+b=(a2 '"b)("(a∧b2)"B. Furthermore, if a"B and c"A, then c.a=c∧a≤a"B as well. As a result, B is a ring ideal of A.

A subset of a Boolean algebra satisfying the two equivalent conditions above is called a Boolean ideal, or an ideal for short. A prime Boolean ideal is a prime lattice ideal, and a maximal Boolean ideal is a maximal lattice ideal. Again, these notions and their ring theoretic counterparts match exactly. In fact, one can say more about these ideals in the case of a Boolean algebra: prime ideals are precisely the maximal ideals. If A is a Boolean ring, and M is a maximal ideal of A, then A/M is isomorphic to {0,1}.

Remark. The dual notion of a Boolean ideal is a Boolean filter, or a filter for short. A Boolean filter is just a lattice filter of the Boolean algebra when considered as a lattice. To see the connection between a Boolean ideal and a Boolean lattice, let us define, for any subset S of a Boolean algebra A, the set S2 :={a2 #"a"S}. It is easy to see that S2 2 =S. Now, if I is an ideal, then I2 is a filter. Conversely, if F is a filter, F2 is a ideal. In fact, given any Boolean algebra, there is a Galois connection

I¦ !I2 ,F¦ !F2

between the set of Boolean ideals and the set of Boolean filters in A . In addition, I is prime iff I2 is. As a result, a filter is prime iff it is an ultrafilter (maximal filter).

BOOLEAN QUOTIENT ALGEBRA

Quotient Algebras via Congruences

Let A be a Boolean algebra. A congruence on A is an equivalence relation Q on A such that Q respects the Boolean operations:

- if aQb and cQd, then (a("c)Q(b("d)
- if aQb, then a2 Qb2

  By de Morgan's laws, we also have aQb and cQd implying $(a\wedge c)Q(b\wedge d)$.

  When a is congruent to b, we usually write aa"b(modQ).

  Let B be the set of congruence classes: B=A/Q, and write [a]Q, or simply [a] for the congruence class containing the element a"A. Define on B the following operations:

- [a] (" [b]:=[a (" b]
- [a]2 :=[a2]

  Because Q respects join and complementation, it is clear that these are well-defined operations on B. Furthermore, we may define $[a]\wedge[b]$:=([a]2 ("[b]2)2 =([a2]("[b2])2 =[a2 ("b2]2 =[(a2 ("b2)2]=$[a\wedge b]$. It is also easy to see that and [0] are the top and bottom elements of B. Finally, it is straightforward to verify that B is a Boolean algebra. The algebra B is called the Boolean quotient algebra of A via the congruence Q.

Quotient Algebras via Ideals and Filters

It is also possible to define quotient algebras via Boolean ideals and Boolean filters. Let A be a Boolean algebra and I an ideal of A. Define binary relation <" on A as follows:

a<"b if and only if a∇b " I,

where ∇ is the symmetric difference operator on A. Then

- <" is an equivalence on A, because
  - o – a∇a=0"I, so <" is reflexive

- o – b∇a=a∇b, so <“ is symmetric, and
- – if a<“b and b<“c, then a<“c; to see this, note that (a”b)(“(b”c)=((a”b)(“b)∧((a”b)(“c2)=(a(“b)∧((a”b)(“c2). Since the LHS (and hence the RHS) is in I, and that a≤a(“b and c2 ≤(a”b)(“c2 , RHS e”a∧c2 =a”c”I too. Similarly c”a”I so that a<“c.
- · 2. <“ respects (“ and 2 , because
- – if a<“b and c<“d, then (a(“c)”(b(“d)=(a(“c)∧(b(“d)2 =(a(“c)∧(b2 ‘”d2)=(a∧(b2 ‘”d2))(“(c∧(b2 ‘”d2))≤(a∧b2)(“(c∧d2)”I, so that (a(“c)”(b(“d)”I as well. That (b(“d)”(a(“c)”I is proved similarly. Hence (a(“c)<“(b(“d).
- – a2 ∇b2 =a∇b, so <“ preserves 2 .

Thus, <“is a congruence on A. The quotient algebra A/<“ is called the quotient algebra of A via the ideal I, and is often denoted by A/I.

From this congruence <“, one can re-capture the ideal: I=[0].

Dually, one can obtain a quotient algebra from a Boolean filter. Specifically, if F is a filter of a Boolean algebra A, define <“ on A as follows:

a<“b   if and only if a”!b “ F,

where ”! is the biconditional operator on A. Then it is easy to show that <“ too is a congruence on A, so that one forms the quotient algebra of A via the filter F, denoted by A/F. Of course, an easier approach to this is to realize that F is a filter of A iff F2 :={a2 #”a”F} is an ideal of A, and the process of forming A/F2 turns out to be identical to A/F.

From <“, the filter F can be recovered: F=.

In fact, given a congruence Q, the congruence class [0]Q is a Boolean ideal and the congruence class Q is a Boolean filter, and that the quotient algebras derived from Q,[0]Q and Q are all the same:

A/Q=A/[0]Q=A/Q.

The Boolean Representation Theorem

In mathematics, Stone's representation theorem for Boolean algebras states that every Boolean algebra is isomorphic to a certain field of sets. The theorem is fundamental to the deeper understanding of Boolean algebra that emerged in the first half of the 20th century. The theorem was first proved by Marshall H. Stone. Stone was led to it by his study of the spectral theory of operators on a Hilbert space.

Stone spaces

Each Boolean algebra B has an associated topological space, denoted here S(B), called its Stone space. The points in S(B) are the ultrafilters on B, or equivalently the homomorphisms from B to the two-element Boolean algebra. The topology on S(B) is generated by a (closed) basis consisting of all sets of the form

$\{x \in S(B)^* b \in x\}$,

where b is an element of B. This is the topology of pointwise convergence of nets of homomorphisms into the two-element Boolean algebra.

For every Boolean algebra B, S(B) is a compact totally disconnected Hausdorff space; such spaces are called Stone spaces (also profinite spaces). Conversely, given any topological space X, the collection of subsets of X that are clopen (both closed and open) is a Boolean algebra.

REPRESENTATION THEOREM

A simple version of Stone's representation theorem states that every Boolean algebra B is isomorphic to the algebra of clopen subsets of its Stone space S(B). The isomorphism sends an element b " B to the set of all ultrafilters that contain b. This is a clopen set because of the choice of topology on S(B) and because B is a Boolean algebra.

Restating the theorem using the language of category theory; the theorem states that there is a duality between the category of Boolean algebras and the category of Stone spaces. This duality means that in addition to the correspondence between Boolean algebras and their Stone spaces, each homomorphism from a Boolean algebra A to a Boolean algebra B corresponds in a natural way to a continuous function from S(B) to S(A). In other words, there is a contravariant functor that gives an equivalence between the categories. This was an early example of a nontrivial duality of categories.

The theorem is a special case of Stone duality, a more general framework for dualities between topological spaces and partially ordered sets.

The proof requires either the axiom of choice or a weakened form of it. Specifically, the theorem is equivalent to the Boolean prime ideal theorem, a weakened choice principle that states that every Boolean algebra has a prime ideal.

An extension of the classical Stone duality to the category of Boolean spaces (= zero-dimensional locally compact Hausdorff spaces) and continuous maps (respectively, perfect maps) was obtained by G. D. Dimov (respectively, by H. P. Doctor).

DUALITY

Duality principle (or principle of duality) is an important property used mainly in proving various theorems available in boolean algebra.

Duality principle states that in a two-valued boolean algebra, the dual of an algebraic expression can be obtained by interchanging all the OR and AND operator and by replacing 1 by 0 and 0 by 1.

The principle of duality in Boolean algebra states that if you have a true Boolean statement (equation) then the dual of this statement (equation) is true. The dual of a boolean statement is found by replacing the statement's symbols with their counterparts. This means a "0" becomes a "1", "1" becomes a "0", "+" becomes a "." and "." becomes a "+".

Here's an example of the principle of duality in Boolean algebra:

Suppose we have the following true Boolean statement (equation):

(*) 1 + 0 = 1

The dual of this statement is:

(**) 0 . 1 = 0

As we can see, the dual of the true Boolean statement (*) is (**). We found (**) by replacing each symbol from (*) with its Boolean counterpart as described above. Clearly, (**) is also a true Boolean statement.

According to this principle, if we have postulates or theorems of Boolean Algebra for one type of operation then that operation can be converted into another type of operation (i.e., AND can be converted to OR and vice-versa) just by interchanging '0 with 1', '1 with 0', '(+) sign with (.) sign' and '(.) sign with (+) sign'. This principle ensures if a theorem is proved using postulates of Boolean algebra, then the dual of this theorem automatically holds and we need not prove it again separately.

Some Boolean expressions and their corresponding duals are given below,

Given Expression	Dual	Given Expression	Dual
0 = 1	1 = 0	A. (A+B) = A	A + A.B = A
0.1 = 0	1 + 0 = 1	AB = A + B	A+B = A.B
A.0 = 0	A + 1 = 1	(A+C) (A +B) = AB + AC	AC + AB = (A+B). (A+C)
A.B = B. A	A + B = B + A	A+B = AB + AB +AB	AB = (A+B).(A+B).(A+B)
A.A = 0	A + A = 1	AB + A + AB = 0	((A+B)).A.(A+B) = 1
A. (B.C) = (A.B). C	A+(B+C) = (A+B) + C		

Reducing Boolean Expressions

Every Boolean expression must be reduced to its simplest form before realizing it because each logic operation in the expression is carried out using hardware. Thus, realizing the simplest expression requires less circuitry hence reduces the cost of the system. Also, it is highly reliable and less complex in nature. For reducing the Boolean expression, we use the axioms and laws of Boolean algebra (see them in our previous article). Some instructions for reducing the given Boolean expression are listed below,

1. Remove all the parenthesis by multiplying all the terms if present.
2. Group all similar terms which are more than one, then remove all other terms by just keeping one.

   Example: ABC + AB +ABC + AB = ABC +ABC + AB +AB = ABC +AB

3. A variable and its negation together in the same term always results in a 0 value, it can be dropped.
   Example: A. BCC + BC = A. 0 + BC = BC
4. Look for the pair of terms that are identical except for one variable which may be missing in one of the terms. The larger term can be dropped.
   Example: ABCD + ABC = ABC(D + 1) = (ABC).1 = ABC
5. Look for the pair of terms that have the same variables, with one or more variables complimented. If a variable in one term of such a variable is complimented while in the second term it is not, then such terms can be combined into a single term with that variable dropped.
   Example
   ABCD + ABCD = ABC(D + D) = (ABC).1 = ABC
   OR
   AB(C+D) + AB(C+D) = AB [(C+D) + (C+D)] = AB. 1 =AB

Steps used in Duality Theorem

Here are some of the main steps used to solve duality theorem in boolean algebra:

- Change each AND operation to an OR operation
- Change each OR operation to an AND operation
- Complement each 1 and 0

Let's take an example to apply all the above three steps to understand about duality principle in a practical way. This example gives you some idea about how to convert or apply duality theorem or principle in any boolean expression.

1+0 = 1

The dual of the above statement will be:

0.1 = 0

As you can see here, we have done the following:

- changed first 1 to 0
- changed OR (+) to AND (.)
- changed first 0 to 1
- changed second 1 to 0

Infinite Distributivity

5.11 m-Completeness ..

Functional Completeness in Digital Logic

A set of operations is said to be functionally complete or universal if and only if every switching function can be expressed by means of operations in it. A set of Boolean functions is functionally complete, if all other Boolean functions can be constructed from this set and a set of input variables are provided, e.g.

- Set A = {+,*,′ (OR, AND, complement) } are functionally complete.
- Set B = {+,′} are functionally complete
- Set C = {*,′} are functionally complete

Post's Functional Completeness Theorem – Important closed classes of functions:

1. T_0 – class of all 0-preserving functions, such as f(0, 0, … , 0) = 0.
2. T_1 – class of all 1-preserving functions, such as f(1, 1, … , 1) = 1.
3. S – class of self-dual functions, such as $f(x_1, \ldots, x_n) = \neg f(\neg x_1, \ldots, \neg x_n)$.
4. M – class of monotonic functions, such as : $\{x_1, \ldots, x_n\} \leq \{x_1, \ldots, x_n\}$, if $x_i \leq y_i$

   if $\{x_1, \ldots, x_n\} \leq \{x_1, \ldots, x_n\}$

   then $f(x_1, \ldots, x_n) \leq f(x_1, \ldots, x_n)$
5. L – class of linear functions, which can be presented as: $f(x_1, \ldots, x_n) = a_0 + a_1 \cdot x_1 + \ldots + a_n \cdot x_n$; a_i {0, 1}.

Theorem – A system of Boolean functions is functionally complete if and only if for each of the five defined classes T_0, T_1, S, M, L, there is a member of F which does not belong to that class.

These are minimal functionally complete operator sets –

One element –

{′!}, {″!}.

Two elements –

$\{\vee,\neg\},\{\wedge,\neg\},\{\rightarrow,\neg\},\{\leftarrow,\neg\},\{\rightarrow,\bot\},\{\leftarrow,\bot\},\{\rightarrow,\nleftrightarrow\},\{\leftarrow,\nleftrightarrow\},$
$\{\rightarrow,\nrightarrow\},\{\rightarrow,\nleftarrow\},\{\leftarrow,\nrightarrow\},\{\leftarrow,\nleftarrow\},\{\nrightarrow,\neg\},\{\nleftarrow,\neg\},\{\nrightarrow,\top\},\{\nleftarrow,\top\},\{\nrightarrow,\leftrightarrow\},\{\nleftarrow,\leftrightarrow\}.$

Three elements –

$$\{\vee,\leftrightarrow,\bot\},\{\vee,\nleftrightarrow,\top\},\{\wedge,\leftrightarrow,\bot\},\{\wedge,\leftrightarrow,\nleftrightarrow\},\{\wedge,\nleftrightarrow,\top\}.$$

Examples on functional Completeness –

- Check if function F(A,B,C) = A′+BC′ is functionally complete?
- Explanation – Let us start by putting all variables as 'A' so it becomes F(A,A,A) = A′+A.A′ = A′—-(i)

  F(B,B,B) = B′+B.B′ = B′—(ii)

Now substitute F(A,A,A) in place of variable 'A' and F(B,B,B) in place of variable 'C'

F(F(A,A,A),B,F(B,B,B)) = (A′)′+B.(B′)′ = A+B—(iii)

from (i) and (ii) complement is derived and from (iii) operator '+' is derived so this function is functionally complete as from above if function contains {+,′} is functionally complete.

- Check if function F(A,B) = A′+B is functionally complete?
- Explanation – Let us start by putting all variables as 'A' so it becomes
  F(A,A) = A′+A = 1—-(i)
  F(B,B) = B′+B = 1—(ii)
  F(A,0) = A′+0 = A′—(iv)
  Now substitute F(A,0) in place of variable 'A'
  F(F(A,0),B) = (A′)′+B = A+B—(iii)
  from (iv) complement is derived and from (iii) operator '+' is derived so this function is functionally complete as from above if function contains {+,′} is partially functionally complete .
- Check if function F(A,B) = A′B is functionally complete?
- Explanation – Let us start by putting all variables as 'A' so it becomes
  F(A,A) = A′.A′ = 0—-(i)
  F(A,0) = A′.0 = 0—(ii)
  F(A,1) = A′.1 = A′—(iv)
  Now substitute F(A,1) in place of variable 'A'
  F(F(A,1),B) = (A′)′*B = A*B—(iii)
  from (iv) complement is derived and from (iii) operator '*' is derived so this function is functionally complete as from above if function contains {*,′} is partially functionally complete .

Note – If the function becomes functionally complete by substituting '0' or '1' then it is known as partially functionally complete.

· Check if function F(A,B) = A′B+AB′ (EX-OR) is functionally complete?

· Explanation – Let us start by putting all variables as 'A' so it becomes

F(A,1) = A′.1 + A.0 = A′—-(i)
F(A′,B) = AB + A′B′–(ii)
F(A′,B′) = AB′ + A′B–(iii)
F(A,B′) = A′B′ + AB—(iv)

So there is no way to get {+,*,′} according to condition. So EX-OR is non functionally complete .

· Consider the operations

f(X, Y, Z) = X′YZ + XY′ + Y′Z′ and g(X2 , Y, Z) = X2 YZ + X2 YZ2 + XY

Which one of the following is correct?

(A) Both {f} and {g} are functionally complete

(B) Only {f} is functionally complete

(C) Only {g} is functionally complete

(D) Neither {f} nor {g} is functionally complete

6

Applications of Boolean Algebra

The usefulness of Boolean algebra comes from the fact that its rules can be shown to apply to logical statements. A logical statement, or proposition, can either be true or false, just as an equation with real numbers can be true or false depending on the value of the variable. In Boolean algebra, however, variables do not represent the values that make a statement true, instead they represent the truth or falsity of the statement. That is, a Boolean variable can only have one of two values. In the context of symbolic logic these values are true and false.

Boolean algebra is also extremely useful in the field of electrical engineering. In particular, by taking the variables to represent values of on and off (or 0 and 1), Boolean algebra is used to design and analyze digital switching circuitry, such as that found in personal computers, pocket calculators, cd players, cellular telephones, and a host of other electronic products.

Two things about Boolean algebra make it a very important form of mathematics for practical applications. First, statements expressed in everyday language (such as "I will be home today") can be converted into mathematical expressions, such as letters and numbers. Second, those symbols generally have only one of two values. The statements above (P and Q), for example, are either true or false. That means they can be expressed in a binary system: true or false; yes or no; 0 or 1.

Boolean algebra has a variety of uses in the real world. It is used in light switches. This basic is significant especially in a large lighting system where the lighting system is interconnected in such a way that it can be controlled using one or more switches. In this case, several lighting bulbs that are interconnected together depend on one switch to control the power supply (switching on and off). Binary mathematics is the number system most often used with computers. Computer systems consist of magnetic cores that can be switched on or switched off. The numbers 0 and 1 are used to represent the two possible states of a magnetic core. Boolean statements can be represented, then, by the numbers 0 and 1 and also by electrical systems that are either on or off. As a result, when engineers design circuitry for personal computers, pocket calculators, compact disc players, cellular telephones, and a host of other electronic products, they apply the principles of Boolean algebra.

Boolean algebra is a useful and clever way to simplify digital circuits used in microprocessors. It can help reduce the cost and raise the speed and efficiency of computers.

Digital circuits are made up of logic gates. Logic gates are a fundamental digital component that preform simple logical calculations. All logic gates have at least one input and exactly one output and can only take two states: 0 (no electrical current), and 1 (electrical current is present).

Multiple logic gates can be connected from output to input, forming complex pathways. When multiple logic gates work together, they can preform complex mathematical calculations, as well as remember data. They can be more than the sum of their parts.

There are seven logic gates:

1. AND gate
2. NAND gate
3. OR gate
4. NOR gate
5. XOR gate
6. XNOR gate
7. NOT gate

Each of these logic gates can preform its own unique logical operations and return an output state, depending on the states of its inputs.

AND

Both inputs of the AND gate must be on in order for the output to be on. Otherwise, the output is off. Hence, input A and input B must be on in order for the output to be on.

OR

The logic of the OR gate is just what you would expect. The output is only off if all of the inputs are off. If at least one input is on, then the output is on.

NOT

The NOT gate is the simplest logic gate. It has only one input and one output. The output is always not the input. For example, suppose the input of a NOT gate is on. Then the output of that same NOT gate must be off. If the input is off, then the output is on.

NAND

The NAND gate is identical to an AND gate with a NOT gate attached to the output. The output is off if and only if both inputs are on. Otherwise, if at least one input is off, then the output is on.

NOR

For the NOR gate, the output can only be on if both inputs are off. Otherwise, the output is off. Like the NAND gate, the NOR gate is essentially an OR gate whose output is connected to a NOT gate.

XOR

The XOR (exclusive OR) gate is the "glorifie≤ version of the OR gate. The output of the XOR gate is on if and only if exactly one of the inputs is on. If both inputs are on, or if both inputs are off, then the output is off.

XNO

The XNOR gate is the inverted version of the XOR gate. Both inputs must hold the same state in order for the output to be on.

Boolean algebra can be used to simplify logic circuits. Here's a simple example of a logic circuit that can be simplified: A double-NOT. Everyone knows that if you invert something twice, then you get what you started with. This fact can help save NOT gate money and decrease computing time.

Simplifying a double-NOT is just one of the things people can do with Boolean Algebra! The switches can be placed in the up or down position in a wiring system but still controlling the same lights. Wiring can be parallel or serial depending on the spacing between them. When wired serially, both switches have to be switched on to light up the bulbs.when in parallel, one or both switches have to be turned on to light up the bulbs. Then regardless of the state of either switch, the other switch can still light the bulbs. These operations bring out the logical operation that is derived from the boolean algebra.

Using of Boolean Algebra

Boolean algebra is commonly used in industries that produce computers, its implementation is realized by the integrated circuits that are in container form, consisting of a large number of logic gates. The communication between the gates is made possible by the use of an external connection of pins that are related to the input and output lines of the individual gate. The commonly used is the printed circuit board because they have metallic strips. The gate networks can be applied in the internal connections after programming has been done to improve the workability of the integrated circuits. This makes it ready to be incorporated into the systems. Once embedded into the computer systems, the computer will be able to carry out arithmetic and logical operations. Algebra enables the machine to determine the value of an output signal (1 or 0) or to save a value into the storage unit.

Boolean algebra has been used to come up with an analysis of flight accidents. The events that lead to failures in complex systems are incorporated with probabilistic and logical tools. The tools make it possible to determine the causes of failures I.e human factors, environmental factors, and equipment failures. It then breaks down and groups the causal agents of the accident using the fault tree method. The faulty tree is used purposely for maintenance tests, risk analysis, and probability maintenance. To identify the cause of a problem, the faulty tree uses a deductive process to show the relationship between the events that added up to the cause of the accident.I.e (undeveloped event, basic event, and intermediate event in that order). The logical event relationship and diagrams are used in flight management to identify the system failures.

Other systems that have applied boolean algebra are search engines. Search engines such as Google and duckduckgo use boolean algebra to enable the users to get data whenever they request or search. The concept of boolean search has been implemented by the use of logical AND, OR, and NOT. The boolean search considers each webpage on the Internet as an element of a set. By doing so, it is easy to retrieve data whenever it is requested.

BOOLEAN ALGEBRA IN THE LIGHTING SYSTEM

The lighting system is very essential in our day to day life, especially in this century. It is through the lighting system that is traced back to boolean algebra that many homes in the world are lit up. The lighting system has enabled us to have the freedom to do what we could not accomplish. This is because with light one can accomplish certain tasks due to increment in working hours. Many companies in the world have made major advancements due to the availability of lighting systems. Major cities are also lit throughout the night. This has led to the growth of the business. The more the business grows, one will increase the working hours.

Boolean algebra helps the development of modern technology2

The use of computers has lessened human labor in most industries. It has enabled the manufacturing of robots who have taken the place of humans. The automation of tasks is made possible by the use of computers. Robots have been used in hospitals to carry out medical research and accurate healthcare solutions. This has reduced the burden for the researchers. Moreover, algebra has been used to develop computers that are used to monitor various businesses to avoid losses. The use of computers has greatly impacted society since it has reduced the time needed to complete a certain task. I.e the use of robots instead of human labor. More efficient medical services have been offered to patients due to the machines' accuracy.

Application Of Boolean Algebra In Computer Science

Logic circuits are analyzed and simplified by Boolean Algebra. Binary numbers are used only in this program. The following are the numbers 0 and 1. Binary Algebra or logical Algebra are other names for it.

In computer science, Boolean Algebra is important because it teaches the basic designs for the circuits that are used in digital computers.

Boolean algebra is used to design and analyze digital switching circuitry, such as those found in personal computers, pocket calculators, cd players, cellular telephones, and a host of other electronic devices, by taking the variables to represent values of on and off (or 0 and 1).

What Is An Example Of Boolean In Computer Science?

Computer programming is frequently based on Boolean algebra. Any expression with a Boolean value is considered a Boolean expression. In Boolean expressions, such as 3 * 5, x * 5, x * y, and Age * 16, the comparisons are Boolean. In a comparison 3 * 5, the result will always be true, since 3 is always less than 5.

What Is Boolean Algebra Computer Science?

George Boole, a computer scientist who is thought to be one of the founding fathers of computer science, invented Boolean algebra. It deals with statements and their Boolean values. A Boolean algebra variable or function takes one of two values: true or false.

What Is Boolean Algebra Commonly Used For?

Computer programming languages are the primary application of Boolean algebra today. Binomial options pricing models use Boolean algebra to determine when an option should be exercised, which is used in finance.

What Is The Use Of Boolean Logic In Computer?

Boolean logic is a type of logic. A program can help make decisions by comparing simple features. In Boolean logic, all values are either True or False, which is a form of algebra. In order to test the selection and iteration conditions that are based on, we use these values.

How Algebra Is Used In Computer Science?

In computer science, algebra is used to develop algorithms and software that work with mathematical objects. In addition to designing formulas for numerical programs and for complete scientific computations, it is also used to design formulas.

What Is The Use Of Boolean Algebra In Real Life?

Our daily lives are dominated by search engines using boolean algebra. A variety of fields are reached through it. Since data can be retrieved easily, it has changed how we work in schools and workplaces. As a result, books no longer have to search for a particular concept.

What Are The Application Of Boolean Algebra In Switching Theory?

Logic networks of this type were represented by a two-valued Boolean algebra (also known as Switching algebra) as a notation. In most cases, "0" and "1" are the two algebraic values, although "T" and "F" are sometimes used to emphasize the relationship between propositional logic and algebra.

WHAT IS AN EXAMPLE OF A BOOLEAN?

In boolean expressions (named for mathematician George Boole), true or false are evaluated. Here are some examples of common language: • My favorite color is pink. I am afraid of computer programming, so I am not interested in it. The book is funny and a great read.

What Is A Boolean In Technology?

In Boolean data, there are only two variables: true and false. Boolean is a classification method for identifying logical truth values and algebraic variables in computer science.

What Is A Boolean Simple Definition?

In Boolean systems, true/false statements are created by using a logical thought system. There are two types of Boolean values: true and false. In Boolean expressions, the operators AND, OR, XOR are used to compare values and not to return a true or false result.

What Is Boolean Logic In Programming?

In Boolean logic, an expression is determined whether it is true or false by using a Boolean function. A computer can use boolean logic to make decisions – it can either make a true statement or a false statement. In addition to variables, boolean logic is the basis for programming as well.

What Is Boolean Used For In Computer Science?

A boolean or bool is a type of data with two possible values: true and false, respectively, in computer science. In modern digital computers, it is called George Boole because it is named after the English mathematician and logician George Boole.

What Is An Example Of Boolean Logic?

A Boolean logic is defined as the use of words and phrases such as "and," "or," and "not" in search tools to get the most relevant results. In Boolean logic, recipes AND potatoes are used to find recipes containing potatoes.

What Is An Example Of A Boolean Operator?

"AND", "OR", and "NOT" are the three Boolean operators. Using them in library databases (between your keywords) can make each search more precise, and save you time.

MINTERM AND MAXTERM EXPANSIONS

So far we have been finding Sum-Of-Product (SOP) solutions to logic reduction problems. For each of these SOP solutions, there is also a Product-Of-Sums solution (POS), which could be more useful, depending on the application.

Before working a Product-Of-Sums solution, we need to introduce some new terminology. The procedure below for mapping product terms is not new to this chapter.

We just want to establish a formal procedure for minterms for comparison to the new procedure for maxterms.

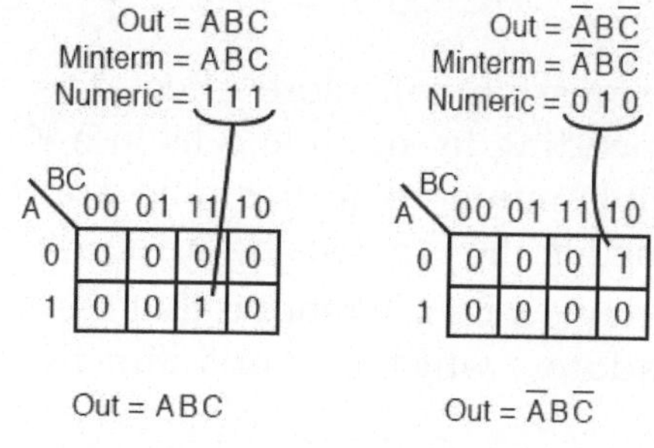

Minterm

A minterm is a Boolean expression resulting in 1 for the output of a single cell, and 0s for all other cells in a Karnaugh map, or truth table. If a minterm has a single 1 and the remaining cells as 0s, it would appear to cover a minimum area of 1s.

The illustration above left shows the minterm ABC, a single product term, as a single 1 in a map that is otherwise 0s. We have not shown the 0s in our Karnaugh maps up to this point, as it is customary to omit them unless specifically needed. Another minterm A′BC′ is shown above right.

The point to review is that the address of the cell corresponds directly to the minterm being mapped. That is, the cell 111 corresponds to the minterm ABC above left.

Above right we see that the minterm A′BC′ corresponds directly to the cell 010. A Boolean expression or map may have multiple minterms.

Referring to the above figure, Let's summarize the procedure for placing a minterm in a K-map:

- Identify the minterm (product term) term to be mapped.
- Write the corresponding binary numeric value.
- Use binary value as an address to place a 1 in the K-map
- Repeat steps for other minterms (P-terms within a Sum-Of-Products).

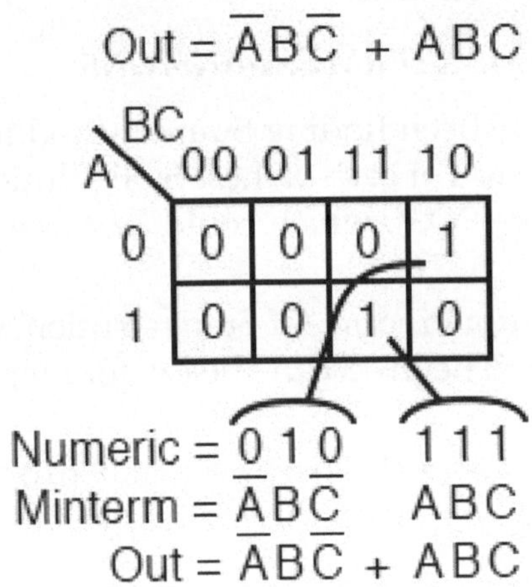

A Boolean expression will more often than not consist of multiple minterms corresponding to multiple cells in a Karnaugh map as shown above. The multiple minterms in this map are the individual minterms which we examined in the previous figure above.

The point we review for reference is that the 1s come out of the K-map as a binary cell address which converts directly to one or more product terms.

By directly we mean that a 0 corresponds to a complemented variable, and a 1 corresponds to a true variable. Example: 010 converts directly to A′BC′.

There was no reduction in this example. Though, we do have a Sum-Of-Products result from the minterms.

Referring to the above figure, Let's summarize the procedure for writing the Sum-Of-Products reduced Boolean equation from a K-map:

- Form largest groups of 1s possible covering all minterms. Groups must be a power of 2.
- Write binary numeric value for groups.
- Convert binary value to a product term.
- Repeat steps for other groups. Each group yields a p-terms within a Sum-Of-Products.

Nothing new so far, a formal procedure has been written down for dealing with minterms. This serves as a pattern for dealing with maxterms.

Next we attack the Boolean function which is 0 for a single cell and 1s for all others.

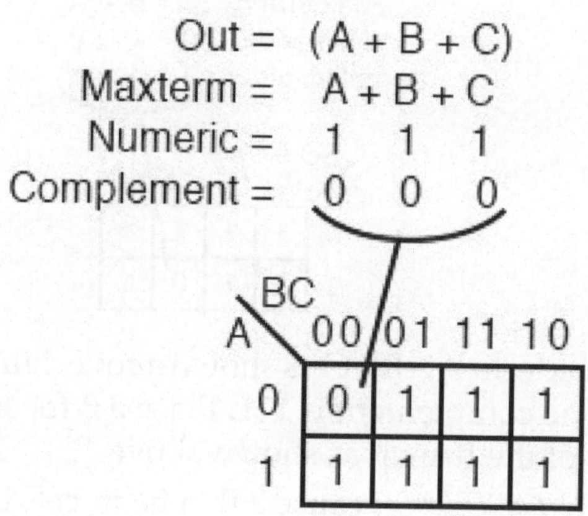

Maxterm

A maxterm is a Boolean expression resulting in a 0 for the output of a single cell expression, and 1s for all other cells in the Karnaugh map, or truth table. The illustration above left shows the maxterm (A+B+C), a single sum term, as a single 0 in a map that is otherwise 1s.

If a maxterm has a single 0 and the remaining cells as 1s, it would appear to cover a maximum area of 1s.

There are some differences now that we are dealing with something new, maxterms. The maxterm is a 0, not a 1 in the Karnaugh map. A maxterm is a sum term, (A+B+C) in our example, not a product term. It also looks strange that (A+B+C) is mapped into the cell 000.

For the equation Out=(A+B+C)=0, all three variables (A, B, C) must individually be equal to 0. Only (0+0+0)=0 will equal 0. Thus we place our sole 0 for minterm (A+B+C) in cell A,B,C=000 in the K-map, where the inputs are all 0 .

This is the only case which will give us a 0 for our maxterm. All other cells contain 1s because any input values other than ((0,0,0) for (A+B+C) yields 1s upon evaluation.

Referring to the above figure, the procedure for placing a maxterm in the K-map is:

- Identify the Sum term to be mapped.
- Write corresponding binary numeric value.
- Form the complement
- Use the complement as an address to place a 0 in the K-map
- Repeat for other maxterms (Sum terms within Product-of-Sums expression).

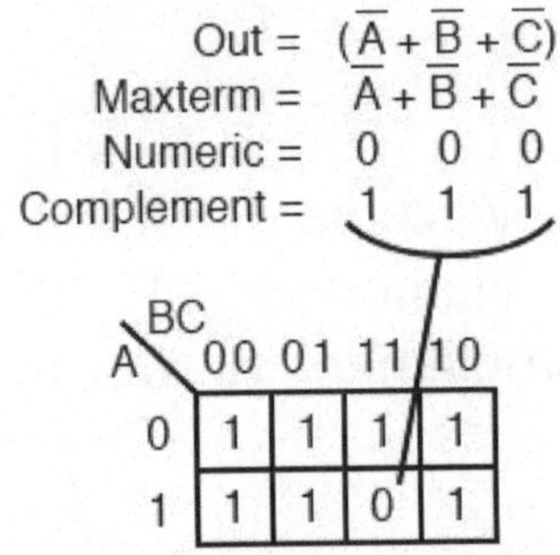

Another maxterm A′+B′+C′ is shown above. Numeric 000 corresponds to A′+B′+C′. The complement is 111. Place a 0 for maxterm (A′+B′+C′) in this cell (1,1,1) of the K-map as shown above.

Why should (A′+B′+C′) cause a 0 to be in cell 111? When A′+B′+C′ is (1′+1′+1′), all 1s in, which is (0+0+0) after taking complements, we have the only condition that will give us a 0. All the 1s are complemented to all 0s, which is 0 when ORed.

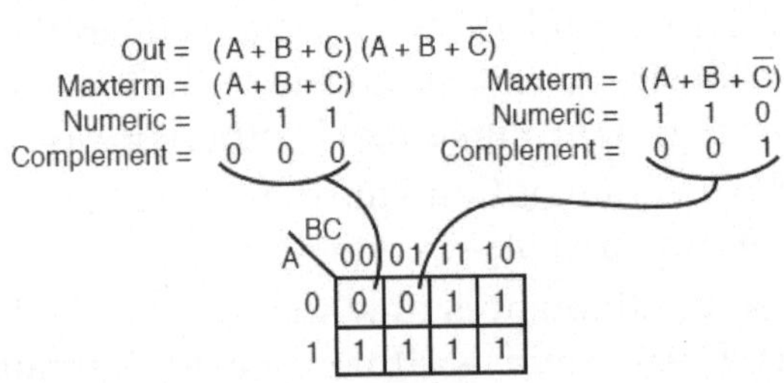

A Boolean Product-Of-Sums expression or map may have multiple maxterms as shown above. Maxterm (A+B+C) yields numeric 111 which complements to 000, placing a 0 in cell (0,0,0). Maxterm (A+B+C') yields numeric 110 which complements to 001, placing a 0 in cell (0,0,1).

Now that we have the k-map setup, what we are really interested in is showing how to write a Product-Of-Sums reduction. Form the 0s into groups. That would be a group of two below. Write the binary value corresponding to the sum-term which is (0,0,X).

Both A and B are 0 for the group. But, C is both 0 and 1 so we write an X as a place holder for C. Form the complement (1,1,X). Write the Sum-term (A+B) discarding the C and the X which held its' place.

In general, expect to have more sum-terms multiplied together in the Product-Of-Sums result. Though, we have a simple example here.

$$\text{Out} = (A + B + C)(A + B + \overline{C})$$

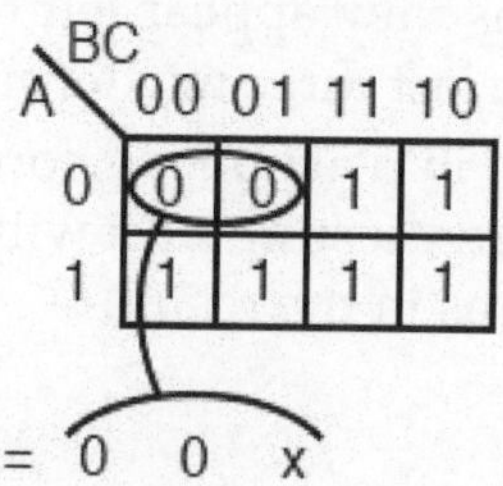

A B C = 0 0 x
Complement = 1 1 x
Sum-term = (A + B)
Out = (A + B)

Let's summarize the procedure for writing the Product-Of-Sums Boolean reduction for a K-map:

- Form largest groups of 0s possible, covering all maxterms. Groups must be a power of 2.
- Write binary numeric value for group.
- Complement binary numeric value for group.
- Convert complement value to a sum-term.
- Repeat steps for other groups. Each group yields a sum-term within a Product-Of-Sums result.

Examples

Example:

Simplify the Product-Of-Sums Boolean expression below, providing a result in POS form.

$$\text{Out} = (A+B+C+\overline{D})(A+B+\overline{C}+D)(A+\overline{B}+C+\overline{D})(A+\overline{B}+\overline{C}+D)$$
$$(\overline{A}+\overline{B}+\overline{C}+D)(\overline{A}+B+C+\overline{D})(\overline{A}+B+\overline{C}+D)$$

Solution:

Transfer the seven maxterms to the map below as 0s. Be sure to complement the input variables in finding the proper cell location.

$$Out = (A+B+C+\overline{D})(A+B+\overline{C}+D)(A+\overline{B}+C+\overline{D})(A+\overline{B}+\overline{C}+D)$$
$$(\overline{A}+\overline{B}+\overline{C}+D)(\overline{A}+B+C+\overline{D})(\overline{A}+B+\overline{C}+D)$$

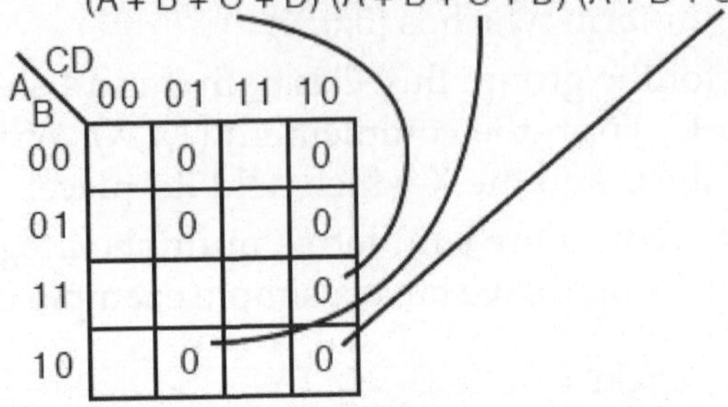

We map the 0s as they appear left to right top to bottom on the map above. We locate the last three maxterms with leader lines..

Once the cells are in place above, form groups of cells as shown below. Larger groups will give a sum-term with fewer inputs. Fewer groups will yield fewer sum-terms in the result.

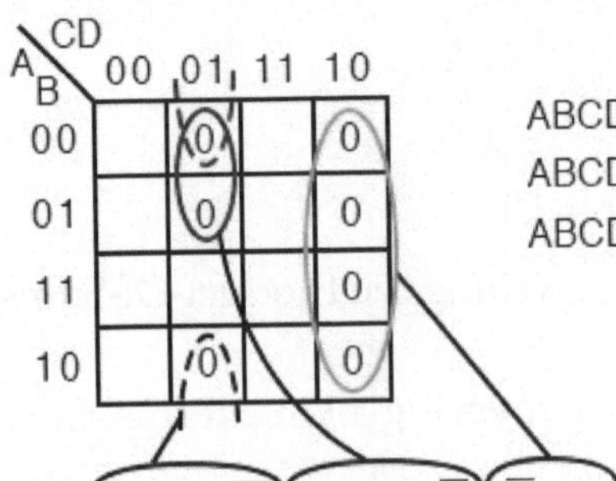

input		complement		Sum-term
ABCD = X001	>	X110	>	$(B+C+\overline{D})$
ABCD = 0X01	>	1X10	>	$(A+C+\overline{D})$
ABCD = XX10	>	XX01	>	$(\overline{C}+D)$

$$Out = (B+C+\overline{D})(A+C+\overline{D})(\overline{C}+D)$$

We have three groups, so we expect to have three sum-terms in our POS result above. The group of 4-cells yields a 2-variable sum-term. The two groups of 2-cells give us two 3-variable sum-terms.

Details are shown for how we arrived at the Sum-terms above. For a group, write the binary group input address, then complement it, converting that to the Boolean sum-term. The final result is product of the three sums.

Boolean Algebra: Multiplying out and Factoring expressions

Multiplication

Consider the following equations:

$0 \times 0 = 0$

$0 \times 1 = 0$

$1 \times 0 = 0$

$1 \times 1 = 1$

As you can see, multiplication behaves in exactly the same way with Boolean algebra as it does in real arithmetic: anything multiplied by zero becomes zero, and anything multiplied by one stays the same.

You should be able to see that the multiplication operator has exactly the same effect as the logical AND operator, and this is why the AND function is represented by multiplication:

A AND B = A . B = A B

Exclusive-OR

Exclusive OR Gate, also known as EX OR Gate or XOR Gate, is an important digital logic gate, which implements an exclusive or logic i.e., the output is HIGH if and only if one of the inputs is HIGH. If both the inputs are LOW or HIGH, then the output is LOW.

XOR Symbol

There are multiple standards for defining an electronic component. Generally, we follow the IEEE (Institute of Electrical and Electronics Engineers) and IEC (International Electrotechnical Commission) standards. The XOR logic symbol in IEEE and IEC standards is shown below.

The Boolean expression for XOR gate cannot determined directly like AND, OR gates. As it is a Hybrid gate, the Boolean expression of output of XOR gate is given by a combining of Multiplication, Addition and inverting of inputs. We have to use the Karnaugh Maps or K – Maps along with the truth table to derive the Boolean Expression of XOR gate.

Equivalance operations

Equivalence statements

Two statements are said to be equivalent if they have the same truth value.

Example

Following are two statements.

p = It is false that he is a singer or he is a dancer.

q = He is not a singer and he is not a dancer.

The first statement p consists of negation of two simple proposition

a = He is a singer.

b = He is a dancer.

They are connected by an OR operator (connective)

so we can write,

p = ~(a (" b)

The second statement q consists of two simple proposition which are negation of a and b

~a = He is not a singer.

~b = He is not a dancer.

They are connected by an AND operator (connective)

so we can write,

q = ~a ‴ ~b

We have,

p = ~(a (" b)

q = ~a ‴ ~b

following are the truth tables for p and q

a	b	a v b	p = ~(a v b)
0	0	0	1
0	1	1	0
1	0	1	0
1	1	1	0

a	b	~a	~b	q = ~a ∧ ~b
0	0	1	1	1
0	1	1	0	0
1	0	0	1	0
1	1	0	0	0

We can see that the truth values are same for both the statements. So, p and q are equivalent statements.

Now we will cover some equivalence laws.

Properties of 0

If x is a statement then,

$0 + x = x$

$0 . x = 0$

where + is the OR operator and

. is the AND operator

Truth table

0	x	0 + x
0	0	0
0	1	1

0 + x = x

0	x	0 . x
0	0	0
0	1	0

0 . x = 0

Properties of 1

If x is a statement then,

$1 + x = 1$

$1 . x = x$

where + is the OR operator and . is the AND operator

Truth table

1	x	1 + x
1	0	1
1	1	1

1 + x = 1

1	x	1 . x
1	0	0
1	1	1

1 . x = x

Involution

If p is a statement then,

$\sim(\sim p) = p$

where ~ is the NOT operator

Truth table

p	~p	~(~p)
0	1	0
1	0	1

we can see that ~(~p) = p

Idempotence Law

If p is a statement then,

$p + p = p$

$p \cdot p = p$

where + is the OR operator and

. is the AND operator

Truth table

p	p	p + p
0	0	0
1	1	1

we can see that p + p = p

Absorption Law

If p and q are two statements then,

p + (p.q) = p

p . (p + q) = p

where + is the OR operator and

. is the AND operator

Truth table

p	q	p.q	p + (p.q)
0	0	0	0
0	1	0	0
1	0	0	1
1	1	1	1

we can see that p + (p.q) = p

Complementarity Law

If p is a statement then,

p + (~p) = 1

p . (~p) = 0

where + is the OR operator,

. is the AND operator and

~ is the NOT operator

Truth table

we can see that p + (~p) = 1

p	~p	p + (~p)
0	1	1
1	0	1

p	~p	p . (~p)
0	1	0
1	0	0

and p . (~p) = 0

Commutative Law

If p and q are two statements then,

p + q = q + p

p . q = q . p

where + is the OR operator and

. is the AND operator

Associative Law

If p, q and r are three statements then,

$(p + q) + r = p + (q + r)$

$(p . q) . r = p . (q . r)$

where + is the OR operator and

. is the AND operator

Distributive Law

If p, q and r are three statements then,

$p . (q + r) = (p . q) + (p . r)$

$p + (q . r) = (p + q) . (p + r)$

$p + (\sim p . q) = p + q$

where + is the OR operator,

. is the AND operator and

~ is the NOT operator

De Morgan's Law

If p and q are two statements then,

$\sim(p + q) = \sim p . \sim q$

$\sim(p . q) = \sim p + \sim q$

where + is the OR operator,

. is the AND operator and

~ is the NOT operator

Truth table

p	q	p + q	~(p + q)	~p	~q	~p . ~q
0	0	0	1	1	1	1
0	1	1	0	1	0	0
1	0	1	0	0	1	0
1	1	1	0	0	0	0

we can see that $\sim(p + q) = \sim p . \sim q$

Conditional Elimination

If p and q are two statements then,

$p \sim q = \sim p + q$

where + is the OR operator and

~ is the NOT operator

Bi-conditional Elimination

If p and q are two statements then,

$p \sim q = (p . q) + (\sim p . \sim q)$

where + is the OR operator,

. is the AND operator and

~ is the NOT operator

THE CONSENSUS THEOREM

In Boolean algebra, the consensus theorem or rule of consensus is the identity:

$$xy \vee \bar{x}z \vee yz = xy \vee \bar{x}z$$

The consensus or resolvent of the terms xy and $\bar{x}z$ is yz. It is the conjunction of all the unique literals of the terms, excluding the literal that appears unnegated in one term and negated in the other. If y includes a term which is negated in z (or vice versa), the consensus term y z is false; in other words, there is no consensus term.

The conjunctive dual of this equation is:

$$(x \vee y)(\bar{x} \vee z)(y \vee z) = (x \vee y)(\bar{x} \vee z)$$

Proof

$$\begin{aligned} xy \vee \bar{x}z \vee yz &= xy \vee \bar{x}z \vee (x \vee \bar{x})yz \\ &= xy \vee \bar{x}z \vee xyz \vee \bar{x}yz \\ &= (xy \vee xyz) \vee (\bar{x}z \vee \bar{x}yz) \\ &= xy(1 \vee z) \vee \bar{x}z(1 \vee y) \\ &= xy \vee \bar{x}z \end{aligned}$$

Consensus

The consensus or consensus term of two conjunctive terms of a disjunction is defined when one term contains the literal a and the other the literal $\bar{a}$, an opposition. The consensus is the conjunction of the two terms, omitting both a and $\bar{a}$, and repeated literals. For example, the consensus of $\bar{x}yz$ and $w\bar{y}z$ is $w\bar{x}z$. The consensus is undefined if there is more than one opposition. For the conjunctive dual of the rule, the consensus y (" z can be derived from (x (" y) and $(\bar{x} \vee z)$ through the resolution inference rule. This shows that the LHS is derivable from the RHS (if A → B then A → AB; replacing A with RHS and B with (y (" z)). The RHS can be derived from the LHS simply through the conjunction elimination inference rule. Since RHS → LHS and LHS → RHS (in propositional calculus), then LHS = RHS (in Boolean algebra).

Applications

In Boolean algebra, repeated consensus is the core of one algorithm for calculating the Blake canonical form of a formula. In digital logic, including the consensus term in a circuit can eliminate race hazards.

Conversion of English sentences to Boolean Equations

Let F(x,y) be the predicate that "Person x and Person y are friends." Let L(x,c) be the predicate that "Person x likes cuisine c."

1. No one has any friends - "x,y¬(F(x,y))
2. Tina is not friends with anyone - "x¬(F(Tina,x))
3. Jacob is not friends with anyone who is a friend of John - "x(F(John,x)→¬F(Jacob,x))
4. Austin is friends with anyone who likes Singaporean food - "x(L(x,Singaporean)→F(Austin,x))
5. Pierre has exactly one friend who likes French food - "x(F(Pierre,x)∧L(x,French)∧"y(F(Pierre,y)∧y `" x→¬L(y,French)))

Combinational Logic Design Using A Truth Table

The following steps are a useful formal approach to combinational problems:

1. Devise a truth table of the independent input variables and the resulting output quantities.
2. Write Boolean algebra statements that describe the truth table.
3. Reduce the Boolean algebra.
4. Mechanize the Boolean statements using the appropriate logic gates.

Consider the truth table that defines the OR gate. Using the lines in this table that yield a true result gives.

Since Q is a two-state variable all other input state combinations must yield a false. If the truth table had more than a single output result, each such result would require a separate equation. An alternative is to write an expression for the false condition.

Boolean Algebra Truth Tables

Boolean Algebra Expressions can be used to construct digital logic truth tables for their respective functions

As well as a standard Boolean Expression, the input and output information of any Logic Gate or circuit can be plotted into a standard table to give a visual representation of the switching function of the system.

The table used to represent the boolean expression of a logic gate function is commonly called a Truth Table. A logic gate truth table shows each possible input combination to the gate or circuit with the resultant output depending upon the combination of these input(s).

For example, consider a single 2-input logic circuit with input variables labelled as A and B. There are "four" possible input combinations or 2^2 of "OFF" and "ON" for the two inputs. However, when dealing with Boolean expressions and especially logic gate truth tables, we do not general use "ON" or "OFF" but instead give them bit values which represent a logic level "1" or a logic level "0" respectively.

Then the four possible combinations of A and B for a 2-input logic gate is given as:

- Input Combination 1. – "OFF" – "OFF" or (0, 0)
- Input Combination 2. – "OFF" – "ON" or (0, 1)
- Input Combination 3. – "ON" – "OFF" or (1, 0)
- Input Combination 4. – "ON" – "ON" or (1, 1)

Therefore, a 3-input logic circuit would have 8 possible input combinations or 2^3 and a 4-input logic circuit would have 16 or 2^4, and so on as the number of inputs increases. Then a logic circuit with "n" number of inputs would have 2^n possible input combinations of both "OFF" and "ON".

So in order to keep things simple to understand, in this tutorial we will only deal with standard 2-input type logic gates, but the principals are still the same for gates with more than two inputs.

Then the Truth tables for a 2-input AND Gate, a 2-input OR Gate and a single input NOT Gate are given as:

2-input AND Gate

For a 2-input AND gate, the output Q is true if BOTH input A "AND" input B are both true, giving the Boolean Expression of: (Q = A and B).

Symbol

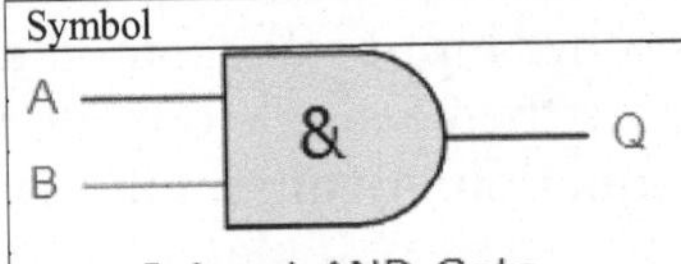

2-input AND Gate

Boolean Expression Q = A.B

Truth Table

A	B	Q
0	0	0
0	1	0
1	0	0
1	1	1

Read as A AND B gives Q

Note that the Boolean Expression for a two input AND gate can be written as: A.B or just simply AB without the decimal point.

2-input OR (Inclusive OR) Gate

For a 2-input OR gate, the output Q is true if EITHER input A "OR" input B is true, giving the Boolean Expression of: (Q = A or B).

Symbol

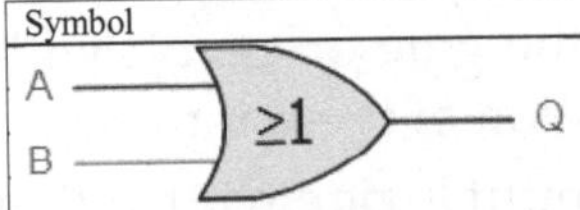

2-input OR Gate

Boolean Expression Q = A+B

Truth Table

A	B	Q
0	0	0
0	1	1
1	0	1
1	1	1

Read as A OR B gives Q

NOT Gate (Inverter)

For a single input NOT gate, the output Q is ONLY true when the input is "NOT" true, the output is the inverse or complement of the input giving the Boolean Expression of: (Q = NOT A).

Symbol

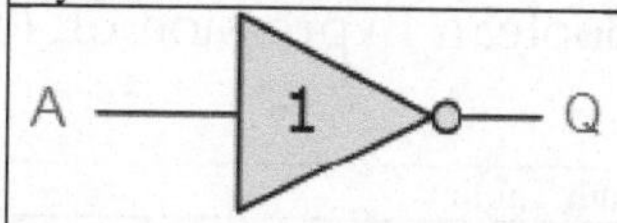

Inverter or NOT Gate

Truth Table	
A	Q
0	1
1	0

Boolean Expression Q = NOT A or A	Read as inversion of A gives Q

The NAND and the NOR Gates are a combination of the AND and OR Gates respectively with that of a NOT Gate (inverter).

2-input NAND (Not AND) Gate

For a 2-input NAND gate, the output Q is NOT true if BOTH input A and input B are true, giving the Boolean Expression of: (Q = not(A AND B)).

Symbol

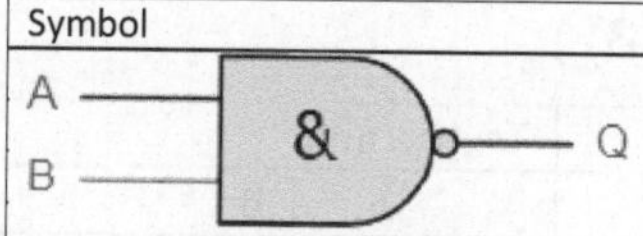

2-input NAND Gate

Truth Table		
A	B	Q
0	0	1
0	1	1
1	0	1
1	1	0

Boolean Expression Q = A .B	Read as A AND B gives NOT-Q

2-input NOR (Not OR) Gate

For a 2-input NOR gate, the output Q is true if BOTH input A and input B are NOT true, giving the Boolean Expression of: (Q = not(A OR B)).

Symbol

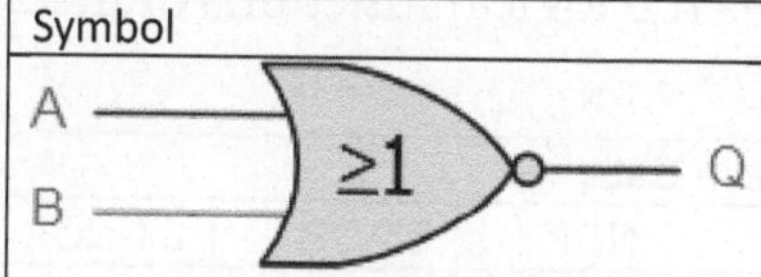

2-input NOR Gate

Truth Table		
A	B	Q
0	0	1
0	1	0
1	0	0
1	1	0

Boolean Expression Q = A+B	Read as A OR B gives NOT-Q

As well as the standard logic gates there are also two special types of logic gate function called an Exclusive-OR Gate and an Exclusive-NOR Gate. The Boolean expression to indicate an Exclusive-OR or Exclusive-NOR function is to a symbol with a plus sign inside a circle, (⊕).

The switching actions of both of these types of gates can be created using the above standard logic gates. However, as they are widely used functions they are now available in standard IC form and have been included here as reference.

2-input EX-OR (Exclusive OR) Gate

For a 2-input Ex-OR gate, the output Q is true if EITHER input A or if input B is true, but NOT both giving the Boolean Expression of: (Q = (A and NOT B) or (NOT A and B)).

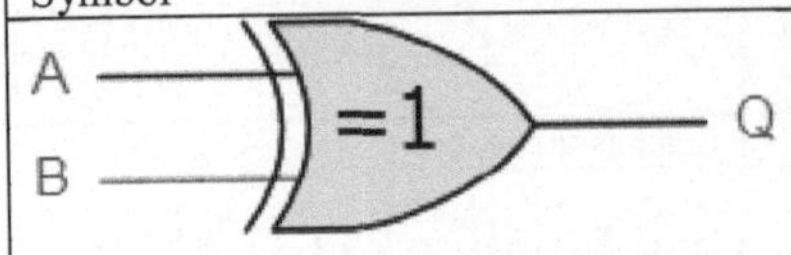

Symbol	Truth Table		
2-input Ex-OR Gate	A	B	Q
	0	0	0
	0	1	1
	1	0	1
	1	1	0
Boolean Expression Q = A ⊕ B			

2-input EX-NOR (Exclusive NOR) Gate

For a 2-input Ex-NOR gate, the output Q is true if BOTH input A and input B are the same, either true or false, giving the Boolean Expression of: (Q = (A and B) or (NOT A and NOT B)).

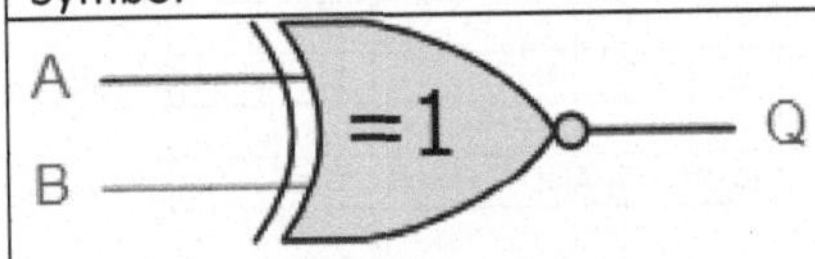

Symbol	Truth Table		
2-input Ex-NOR Gate	A	B	Q
	0	0	1
	0	1	0
	1	0	0
	1	1	1
Boolean Expression Q = A ⊕ B			

Summary of 2-input Logic Gates

The following Truth Table compares the logical functions of the 2-input logic gates above.

Inputs		Truth Table Outputs For Each Gate					
A	B	AND	NAND	OR	NOR	EX-OR	EX-NOR
0	0	0	1	0	1	0	1
0	1	0	1	1	0	1	0
1	0	0	1	1	0	1	0
1	1	1	0	1	0	0	1

The following table gives a list of the common logic functions and their equivalent Boolean notation.

Logic Function	Boolean Notation
AND	A.B
OR	A+B
NOT	A
NAND	A .B
NOR	A+B
EX-OR	(A.B) + (A.B) or A ⊕ B
EX-NOR	(A.B) + (A.B) or A ⊕ B

2-input logic gate truth tables are given here as examples of the operation of each logic function, but there are many more logic gates with 3, 4 even 8 individual inputs. The multiple input gates are no different to the simple 2-input gates above, So a 4-input AND gate would still require ALL 4-inputs to be present to produce the required output at Q and its larger truth table would reflect that.

BINARY SUBTRACTION CIRCUITS

Another basic arithmetic operation to be performed by Digital Computers is the Subtraction. Subtraction is a mathematical operation in which one integer number is deducted from another to obtain the equivalent quantity. The number from which other number is to be deducted is called as 'Minuend' and the number subtracted from the minuend is called 'Subtrahend'.

Similar to the binary addition, binary subtraction is also has four possible basic operations. They are:

- 0 – 0 = 0
- 0 – 1 = (Borrow)1 1
- 1 – 0 = 1
- 1 – 1 = 0

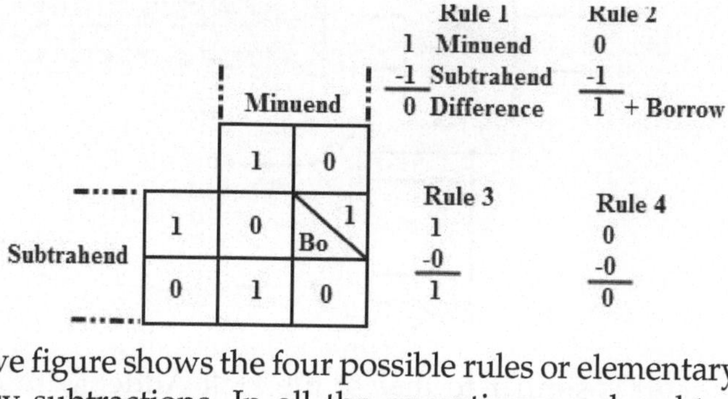

The above figure shows the four possible rules or elementary operations of the binary subtractions. In all the operations, each subtrahend bit is deducted from the minuend bit.

But in the second rule, minuend bit is smaller than the subtrahend bit, hence 1 is borrowed to perform the subtraction. Similar to the adder circuits, basic subtraction circuits are also of two types:

- Half Subtractor
- Full Subtractor

Half Subtractors

A Half Subtractor is a multiple output Combinational Logic Circuit that does the subtraction of two 1-bit binary numbers. It has two inputs and two outputs. The two inputs correspond to the two 1-bit binary numbers and the two outputs corresponds to the Difference bit and Borrow bit (in contrast to Sum and Carry in Half Adder).

The following image shows the block diagram of a Half Subtractor.

Image

Following table shows the truth table of a Half Subtractor.

INPUT		OUTPUT	
A	B	Difference	Borrow
0	0	0	0
0	1	1	1
1	0	1	0
1	1	0	0

From the above truth table, we can say that the 'Difference' output of the Half Subtractor is similar to an XOR output (which is also same as the Sum output of the Half Adder). Thus, the Half Subtraction is also performed by the Ex-OR gate with an AND gate with one inverted input and one normal input, requiring to perform the Borrow operation.

The following image shows the logic circuit of a Half Adder.

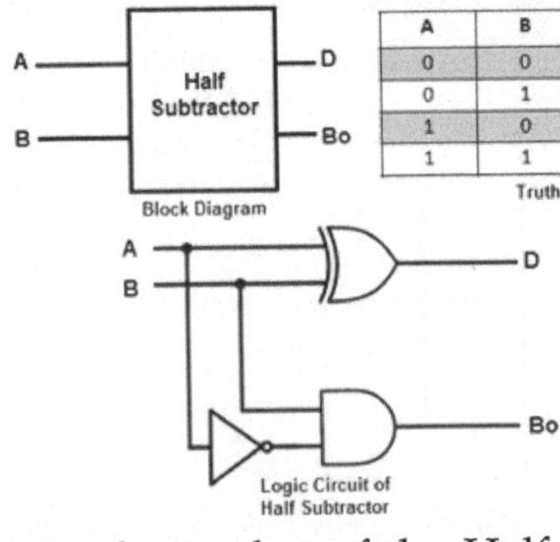

A	B	D	B_o
0	0	0	0
0	1	1	1
1	0	1	0
1	1	0	0

Truth Table

This circuit is similar to that of the Half Adder with only difference being the minuend input i.e., A is complemented before applied at the AND gate to implement the borrow output. In case of multi-digit subtraction, subtraction between the two digits must be performed along with borrow of the previous digit subtraction, and hence a subtractor needs to have three inputs, which is not possible with a Half Subtractor. Therefore, a half subtractor has limited set of applications and strictly speaking, it is not used in practice.

Full Subtractor

A Full Subtractor is a combinational logic circuit which performs a subtraction between the two 1-bit binary numbers and it also considers the borrow of the previous bit i.e., whether 1 has been borrowed by the previous minuend bit.

So, a Full Subtractor has three inputs, in which two inputs corresponding to the two bits to be subtracted (minuend A and subtrahend B), and a borrow bit, usually represented as B_{IN}, corresponding to the borrow operation. There are two outputs, one corresponds to the difference D output and the other Borrow output B_O.

The following table shows the truth table of a Full Subtractor.

INPUT			OUTPUT	
A	B	B_{IN}	D	B_{OUT}
0	0	0	0	0
0	0	1	1	1
0	1	0	1	1
0	1	1	0	1
1	0	0	1	0
1	0	1	0	0
1	1	0	0	0
1	1	1	1	1

By deriving the Boolean expression for the full subtractor from above truth table, we get the expression that tells that a full subtractor can be implemented with half subtractors with OR gate as shown in figure below.

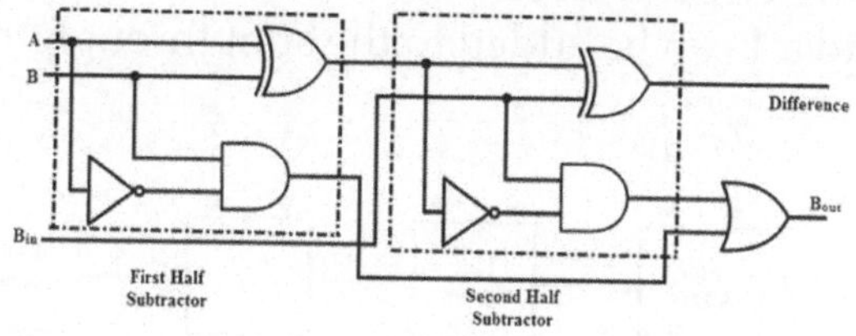

By comparing the adder and subtractor circuits and truth tables, we can observe that the output D in the full subtractor is exactly same as the output S of the full adder. And the only difference is that input variable A is complemented in the full subtractor.

Therefore, it is possible to convert the full adder circuit into full subtractor by simply complementing the input A before it is applied to the gates to produce the final borrow bit output Bo.

Parallel Binary Subtractors

To perform the subtraction of binary numbers with more than one bit, we have to use the Parallel Subtractors. This parallel subtractor can be designed in several ways, including combination of half and full subtractors, all full subtractors, all full adders with subtrahend complement input, etc.

The below figure shows a 4 bit Parallel Binary Subtractor formed by connecting one half subtractor and three full subtractors.

In this subtractor, 4 bit minuend 'A_3 A_2 A_1 A_0' is subtracted by 4 bit subtrahend 'B_3 B_2 B_1 B_0' and the result is the difference output 'D_3 D_2 D_1 D_0'. The borrow output of each subtractor is connected as the borrow input to the next subtractor.

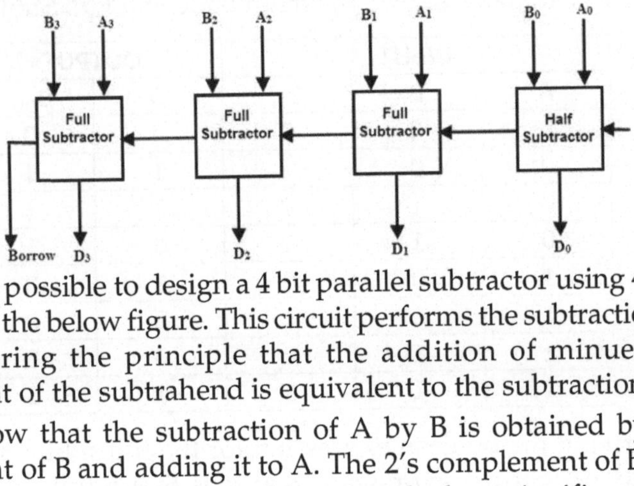

It is also possible to design a 4 bit parallel subtractor using 4 full adders as shown in the below figure. This circuit performs the subtraction operation by considering the principle that the addition of minuend and the complement of the subtrahend is equivalent to the subtraction process.

We know that the subtraction of A by B is obtained by taking 2's complement of B and adding it to A. The 2's complement of B is obtained by taking 1's complement and adding 1 to the least significant pair of bits.

Hence, in this circuit 1's complement of B is obtained with the inverters (NOT gate) and a 1 can be added to the sum through the input carry.

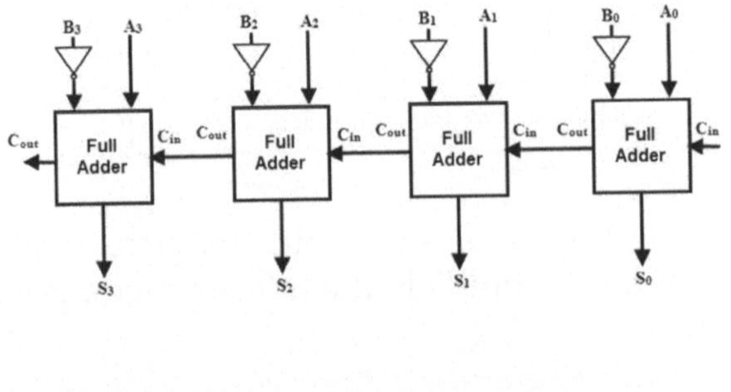

Parallel Adder / Subtractor

The operations of both addition and subtraction can be performed by a one common binary adder. Such binary circuit can be designed by adding an Ex-OR gate with each full adder as shown in below figure. The figure below shows the 4 bit parallel binary adder/subtractor which has two 4 bit inputs as '$A_3\, A_2\, A_1\, A_0$' and '$B_3\, B_2\, B_1\, B_0$'.

The mode input control line M is connected with carry input of the least significant bit of the full adder. This control line decides the type of operation, whether addition or subtraction.

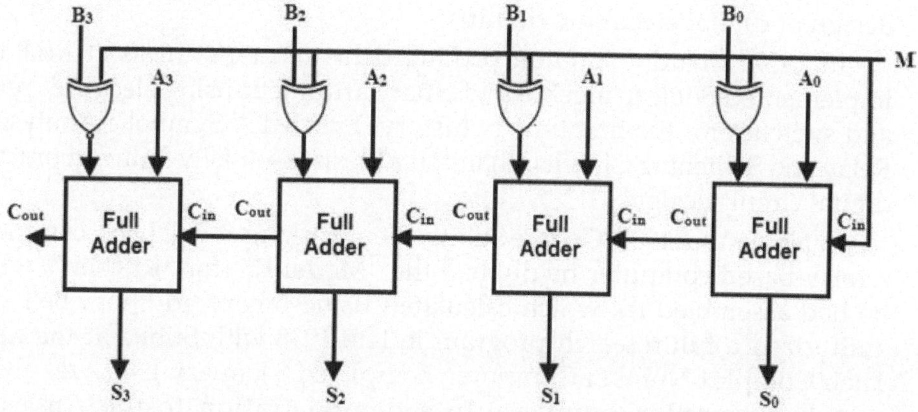

When M= 1, the circuit is a subtractor and when M=0, the circuit becomes adder. The Ex-OR gate consists of two inputs to which one is connected to the B and other to input M. When M = 0, B Ex-OR of 0 produce B. Then, full adders add the B with A with carry input zero and hence an addition operation is performed.

When M = 1, B Ex-OR of 0 produce B complement and also carry input is 1. Hence the complemented B inputs are added to A and 1 is added through the input carry, nothing but a 2's complement operation. Therefore, the subtraction operation is performed.

A complete beginner's tutorial on Binary Adders and Subtractors. You learned different binary adder circuits like Half Adder, Full Adder, Parallel Adder and different binary subtractors like Half Subtractor, Full Subtractor, Parallel Subtractor and also a Combination Parallel Adder / Subtractor circuit.

Binary number

A binary number is a number expressed in the base-2 numeral system or binary numeral system, a method of mathematical expression which uses only two symbols: typically "0" (zero) and "1" (one).

The base-2 numeral system is a positional notation with a radix of 2. Each digit is referred to as a bit, or binary digit. Because of its straightforward implementation in digital electronic circuitry using logic gates, the binary system is used by almost all modern computers and computer-based devices, as a preferred system of use, over various other human techniques of communication, because of the simplicity of the language and the noise immunity in physical implemention.

In 1854, British mathematician George Boole published a landmark paper detailing an algebraic system of logic that would become known as Boolean algebra. His logical calculus was to become instrumental in the design of digital electronic circuitry.

In 1937, Claude Shannon produced his master's thesis at MIT that implemented Boolean algebra and binary arithmetic using electronic relays and switches for the first time in history. Entitled A Symbolic Analysis of Relay and Switching Circuits, Shannon's thesis essentially founded practical digital circuit design.

In November 1937, George Stibitz, then working at Bell Labs, completed a relay-based computer he dubbed the "Model K" (for "Kitchen", where he had assembled it), which calculated using binary addition. Bell Labs authorized a full research program in late 1938 with Stibitz at the helm. Their Complex Number Computer, completed 8 January 1940, was able to calculate complex numbers. In a demonstration to the American Mathematical Society conference at Dartmouth College on 11 September 1940, Stibitz was able to send the Complex Number Calculator remote commands over telephone lines by a teletype. It was the first computing machine ever used remotely over a phone line. Some participants of the conference who witnessed the demonstration were John von Neumann, John Mauchly and Norbert Wiener, who wrote about it in his memoirs.

The Z1 computer, which was designed and built by Konrad Zuse between 1935 and 1938, used Boolean logic and binary floating point numbers.

Representation

Any number can be represented by a sequence of bits (binary digits), which in turn may be represented by any mechanism capable of being in two mutually exclusive states. Any of the following rows of symbols can be interpreted as the binary numeric value of 667:

1	0	1	0	0	1	1	0	1	1
|	—	|	—	—	|	|	—	|	|
☒	☐	☒	☐	☐	☒	☒	☐	☒	☒
y	n	y	n	n	y	y	n	y	y

The numeric value represented in each case is dependent upon the value assigned to each symbol. In the earlier days of computing, switches, punched holes and punched paper tapes were used to represent binary values. In a modern computer, the numeric values may be represented by two different voltages; on a magnetic disk, magnetic polarities may be used. A "positive", "yes", or "on" state is not necessarily equivalent to the numerical value of one; it depends on the architecture in use.

In keeping with customary representation of numerals using Arabic numerals, binary numbers are commonly written using the symbols 0 and 1. When written, binary numerals are often subscripted, prefixed or suffixed in order to indicate their base, or radix. The following notations are equivalent:

- 100101 binary (explicit statement of format)
- 100101b (a suffix indicating binary format; also known as Intel convention)
- 100101B (a suffix indicating binary format)
- bin 100101 (a prefix indicating binary format)
- 100101_2 (a subscript indicating base-2 (binary) notation)
- %100101 (a prefix indicating binary format; also known as Motorola convention)
- 0b100101 (a prefix indicating binary format, common in programming languages)
- 6b100101 (a prefix indicating number of bits in binary format, common in programming languages)
- #b100101 (a prefix indicating binary format, common in Lisp programming languages)

When spoken, binary numerals are usually read digit-by-digit, in order to distinguish them from decimal numerals. For example, the binary numeral 100 is pronounced one zero zero, rather than one hundred, to make its binary nature explicit, and for purposes of correctness. Since the binary numeral 100 represents the value four, it would be confusing to refer to the numeral as one hundred (a word that represents a completely different value, or amount). Alternatively, the binary numeral 100 can be read out as "four" (the correct value), but this does not make its binary nature explicit.

Counting in binary

Counting in binary is similar to counting in any other number system. Beginning with a single digit, counting proceeds through each symbol, in increasing order. Before examining binary counting, it is useful to briefly discuss the more familiar decimal counting system as a frame of reference.

Decimal number	Binary number
0	0
1	1
2	10
3	11
4	100
5	101
6	110
7	111
8	1000
9	1001
10	1010
11	1011
12	1100
13	1101
14	1110
15	1111

Decimal counting

Decimal counting uses the ten symbols 0 through 9. Counting begins with the incremental substitution of the least significant digit (rightmost digit) which is often called the first digit. When the available symbols for this position are exhausted, the least significant digit is reset to 0, and the next digit of higher significance (one position to the left) is incremented (overflow), and incremental substitution of the low-order digit resumes. This method of reset and overflow is repeated for each digit of significance. Counting progresses as follows:

000, 001, 002, ... 007, 008, 009, (rightmost digit is reset to zero, and the digit to its left is incremented)

010, 011, 012, ...

090, 091, 092, ... 097, 098, 099, (rightmost two digits are reset to zeroes, and next digit is incremented)

100, 101, 102, ...

Binary counting

Binary counting follows the exact same procedure, and again the incremental substitution begins with the least significant digit, or bit (the rightmost one, also called the first bit), except that only the two symbols 0 and 1 are available. Thus, after a bit reaches 1 in binary, an increment resets it to 0 but also causes an increment of the next bit to the left:

0000,

0001, (rightmost bit starts over, and next digit is incremented)

0010, 0011, (rightmost two bits start over, and next bit is incremented)

0100, 0101, 0110, 0111, (rightmost three bits start over, and the next bit is incremented)

1000, 1001, 1010, 1011, 1100, 1101, 1110, 1111 ...

In the binary system, each bit represents an increasing power of 2, with the rightmost bit representing 2^0, the next representing 2^1, then 2^2, and so on. The value of a binary number is the sum of the powers of 2 represented by each "1"/ bit. For example, the binary number 100101 is converted to decimal form as follows:

$100101_2 = [(1) \times 2^5] + [(0) \times 2^4] + [(0) \times 2^3] + [(1) \times 2^2] + [(0) \times 2^1] + [(1) \times 2^0]$

$100101_2 = [1 \times 32] + [0 \times 16] + [0 \times 8] + [1 \times 4] + [0 \times 2] + [1 \times 1]$

$100101_2 = 37_{10}$

Fractions

Fractions in binary arithmetic terminate only if 2 is the only prime factor in the denominator. As a result, 1/10 does not have a finite binary representation (10 has prime factors 2 and 5). This causes 10 × 0.1 not to precisely equal 1 in floating-point arithmetic. As an example, to interpret the binary expression for 1/3 = .010101..., this means: $1/3 = 0 \times 2^{-1} + 1 \times 2^{-2} + 0 \times 2^{-3} + 1 \times 2^{-4} + \ldots = 0.3125 + \ldots$ An exact value cannot be found with a sum of a finite number of inverse powers of two, the zeros and ones in the binary representation of 1/3 alternate forever.

Fraction	Decimal	Binary	Fractional approximation
1/1	1 or 0.999...	1 or 0.111...	1/2 + 1/4 + 1/8...
1/2	0.5 or 0.4999...	0.1 or 0.0111...	1/4 + 1/8 + 1/16 . . .
1/3	0.333...	0.010101...	1/4 + 1/16 + 1/64 . . .
1/4	0.25 or 0.24999...	0.01 or 0.00111...	1/8 + 1/16 + 1/32 . . .
1/5	0.2 or 0.1999...	0.00110011...	1/8 + 1/16 + 1/128 . . .
1/6	0.1666...	0.0010101...	1/8 + 1/32 + 1/128 . . .
1/7	0.142857142857...	0.001001...	1/8 + 1/64 + 1/512 . . .
1/8	0.125 or 0.124999...	0.001 or 0.000111...	1/16 + 1/32 + 1/64 . . .
1/9	0.111...	0.000111000111...	1/16 + 1/32 + 1/64 . . .
1/10	0.1 or 0.0999...	0.000110011...	1/16 + 1/32 + 1/256 . . .
1/11	0.090909...	0.00010111010001011101...	1/16 + 1/64 + 1/128 . . .
1/12	0.08333...	0.00010101...	1/16 + 1/64 + 1/256 . . .
1/13	0.076923076923...	0.000100111011000100111011...	1/16 + 1/128 + 1/256 . . .
1/14	0.0714285714285...	0.0001001001...	1/16 + 1/128 + 1/1024 . . .
1/15	0.0666...	0.00010001...	1/16 + 1/256 . . .
1/16	0.0625 or 0.0624999...	0.0001 or 0.0000111...	1/32 + 1/64 + 1/128 . . .

Binary arithmetic

Arithmetic in binary is much like arithmetic in other numeral systems. Addition, subtraction, multiplication, and division can be performed on binary numerals.

Addition

The simplest arithmetic operation in binary is addition. Adding two single-digit binary numbers is relatively simple, using a form of carrying:

$0 + 0 \rightarrow 0$

$0 + 1 \rightarrow 1$

$1 + 0 \rightarrow 1$

$1 + 1 \rightarrow 0$, carry 1 (since $1 + 1 = 2 = 0 + (1 \times 2^1)$)

Adding two "1" digits produces a digit "0", while 1 will have to be added to the next column. This is similar to what happens in decimal when certain single-digit numbers are added together; if the result equals or exceeds the value of the radix (10), the digit to the left is incremented:

$5 + 5 \rightarrow 0$, carry 1 (since $5 + 5 = 10 = 0 + (1 \times 10^1)$)

$7 + 9 \rightarrow 6$, carry 1 (since $7 + 9 = 16 = 6 + (1 \times 10^1)$)

This is known as carrying. When the result of an addition exceeds the value of a digit, the procedure is to "carry" the excess amount divided by the radix (that is, 10/10) to the left, adding it to the next positional value. This is correct since the next position has a weight that is higher by a factor equal to the radix. Carrying works the same way in binary:

```
 1 1 1 1 1   (carried digits)
   0 1 1 0 1
 + 1 0 1 1 1
 ___________
= 1 0 0 1 0 0 = 36
```

In this example, two numerals are being added together: 01101_2 (13_{10}) and 10111_2 (23_{10}). The top row shows the carry bits used. Starting in the rightmost column, $1 + 1 = 10_2$. The 1 is carried to the left, and the 0 is written at the bottom of the rightmost column. The second column from the right is added: $1 + 0 + 1 = 10_2$ again; the 1 is carried, and 0 is written at the bottom. The third column: $1 + 1 + 1 = 11_2$. This time, a 1 is carried, and a 1 is written in the bottom row. Proceeding like this gives the final answer 100100_2 (36_{10}).

When computers must add two numbers, the rule that: x xor y = (x + y) mod 2 for any two bits x and y allows for very fast calculation, as well.

Long carry method

A simplification for many binary addition problems is the Long Carry Method or Brookhouse Method of Binary Addition. This method is generally useful in any binary addition in which one of the numbers contains a long "string" of ones. It is based on the simple premise that under the binary system, when given a "string" of digits composed entirely of n ones (where n is any integer length), adding 1 will result in the number 1 followed by a string of n zeros. That concept follows, logically, just as in the decimal system, where adding 1 to a string of n 9s will result in the number 1 followed by a string of n 0s:

Binary		Decimal
11111	likewise	99999
+ 1		+ 1
100000		100000

Such long strings are quite common in the binary system. From that one finds that large binary numbers can be added using two simple steps, without excessive carry operations. In the following example, two numerals are being added together: $1\,1\,1\,0\,1\,1\,1\,1\,1\,0_2$ (958_{10}) and $1\,0\,1\,0\,1\,1\,0\,0\,1\,1_2$ (691_{10}), using the traditional carry method on the left, and the long carry method on the right:

Traditional Carry Method vs. Long Carry Method

111 11111 (carried digits) 1 ! 1 ! carry the 1 until it is one digit past the "string" below

1110111110 ~~111~~0~~11111~~0 cross out the "string",

+ 1010110011 + 10~~1~~01100~~1~~1 and cross out the digit that was added to it

=11001110001 11001110001

The top row shows the carry bits used. Instead of the standard carry from one column to the next, the lowest-ordered "1" with a "1" in the corresponding place value beneath it may be added and a "1" may be carried to one digit past the end of the series. The "use≤ numbers must be crossed off, since they are already added. Other long strings may likewise be cancelled using the same technique. Then, simply add together any remaining digits normally. Proceeding in this manner gives the final answer of $1\,1\,0\,0\,1\,1\,1\,0\,0\,0\,1_2$ (1649_{10}). In our simple example using small numbers, the traditional carry method required eight carry operations, yet the long carry method required only two, representing a substantial reduction of effort.

Addition table

	0	1
0	0	1
1	1	10

The binary addition table is similar, but not the same, as the truth table of the logical disjunction operation (" . The difference is that 1 (" 1 = 1, while 1 + 1 = 10.

Subtraction

Subtraction works in much the same way:

0 " 0 → 0

0 " 1 → 1, borrow 1

1 " 0 → 1

1 " 1 → 0

Subtracting a "1" digit from a "0" digit produces the digit "1", while 1 will have to be subtracted from the next column. This is known as borrowing. The principle is the same as for carrying. When the result of a subtraction is less than 0, the least possible value of a digit, the procedure is to "borrow" the deficit divided by the radix (that is, 10/10) from the left, subtracting it from the next positional value.

```
  *  ***  (starred columns are borrowed from)
 1101110
"  10111
___________
=1010111
  *        (starred columns are borrowed from)
 1011111
- 101011
___________
=0110100
```

Subtracting a positive number is equivalent to adding a negative number of equal absolute value. Computers use signed number representations to handle negative numbers—most commonly the two's complement notation. Such representations eliminate the need for a separate "subtract" operation. Using two's complement notation subtraction can be summarized by the following formula:

A ″ B = A + not B + 1

Multiplication

Multiplication in binary is similar to its decimal counterpart. Two numbers A and B can be multiplied by partial products: for each digit in B, the product of that digit in A is calculated and written on a new line, shifted leftward so that its rightmost digit lines up with the digit in B that was used. The sum of all these partial products gives the final result.

Since there are only two digits in binary, there are only two possible outcomes of each partial multiplication:

- If the digit in B is 0, the partial product is also 0
- If the digit in B is 1, the partial product is equal to A

For example, the binary numbers 1011 and 1010 are multiplied as follows:

```
      1 0 1 1  (A)
    × 1 0 1 0  (B)
    ---------
      0 0 0 0  ! Corresponds to the rightmost 'zero' in B
 +   1 0 1 1   ! Corresponds to the next 'one' in B
 +  0 0 0 0
 + 1 0 1 1
 ---------------
 = 1 1 0 1 1 1 0
```

Binary numbers can also be multiplied with bits after a binary point:

```
         1 0 1 . 1 0 1   A (5.625 in decimal)
       × 1 1 0 . 0 1     B (6.25 in decimal)
       -----------------
           1 . 0 1 1 0 1   ! Corresponds to a 'one' in B
 +        0 0 . 0 0 0 0    ! Corresponds to a 'zero' in B
 +       0 0 0 . 0 0 0
 +      1 0 1 1 . 0 1
 +     1 0 1 1 0 . 1
 ---------------------------
 =   1 0 0 0 1 1 . 0 0 1 0 1 (35.15625 in decimal)
```

See also Booth's multiplication algorithm.

Multiplication table

	0	1
0	0	0
1	0	1

The binary multiplication table is the same as the truth table of the logical conjunction operation ʹʹʹ.

Bitwise operations

Though not directly related to the numerical interpretation of binary symbols, sequences of bits may be manipulated using Boolean logical operators. When a string of binary symbols is manipulated in this way, it is called a bitwise operation; the logical operators AND, OR, and XOR may be performed on corresponding bits in two binary numerals provided as input.

The logical NOT operation may be performed on individual bits in a single binary numeral provided as input. Sometimes, such operations may be used as arithmetic short-cuts, and may have other computational benefits as well. For example, an arithmetic shift left of a binary number is the equivalent of multiplication by a (positive, integral) power of 2.

Bibliography

- Boole, George (2003) [1854]. An Investigation of the Laws of Thought. Prometheus Books. ISBN 978-1-59102-089-9.
- "The name Boolean algebra (or Boolean 'algebras') for the calculus originated by Boole, extended by Schröder, and perfected by Whitehead seems to have been first suggested by Sheffer, in 1913." E. V. Huntington, "New sets of independent postulates for the algebra of logic, with special reference to Whitehead and Russell's Principia mathematica", in Trans. Amer. Math. Soc. 35 (1933), 274-304; footnote, page 278.
- Peirce, Charles S. (1931). Collected Papers. Vol. 3. Harvard University Press. p. 13. ISBN 978-0-674-13801-8.
- Givant, Steven; Halmos, Paul (2009). Introduction to Boolean Algebras. Undergraduate Texts in Mathematics, Springer. ISBN 978-0-387-40293-2.
- J. Michael Dunn; Gary M. Hardegree (2001). Algebraic methods in philosophical logic. Oxford University Press US. p. 2. ISBN 978-0-19-853192-0.
- Weisstein, Eric W. "Boolean Algebra". mathworld.wolfram.com. Retrieved 2020-09-02.
- Norman Balabanian; Bradley Carlson (2001). Digital logic design principles. John Wiley. pp. 39–40. ISBN 978-0-471-29351-4., online sample
- Rajaraman & Radhakrishnan (2008-03-01). Introduction To Digital Computer Design. PHI Learning Pvt. Ltd. p. 65. ISBN 978-81-203-3409-0.
- Paul Halmos, and Givant, Steven (1998) Logic as Algebra. Dolciani Mathematical Expositions No. 21. The Mathematical Association of America.

- Johnstone, Peter T. (1982) Stone Spaces. Cambridge University Press. ISBN 0-521-23893-5.
- Burris, Stanley N., and H. P. Sankappanavar, H. P.(1981) A Course in Universal Algebra. Springer-Verlag. ISBN 3-540-90578-2.
- John A. Camara (2010). Electrical and Electronics Reference Manual for the Electrical and Computer PE Exam. www.ppi2pass.com. p. 41. ISBN 978-1-59126-166-7.
- Shin-ichi Minato, Saburo Muroga (2007). "Binary Decision Diagrams". In Wai-Kai Chen (ed.). The VLSI handbook (2nd ed.). CRC Press. ISBN 978-0-8493-4199-1. chapter 29.
- Alan Parkes (2002). Introduction to languages, machines and logic: computable languages, abstract machines and formal logic. Springer. p. 276. ISBN 978-1-85233-464-2.
- Jon Barwise; John Etchemendy; Gerard Allwein; Dave Barker-Plummer; Albert Liu (1999). Language, proof, and logic. CSLI Publications. ISBN 978-1-889119-08-3.
- Ben Goertzel (1994). Chaotic logic: language, thought, and reality from the perspective of complex systems science. Springer. p. 48. ISBN 978-0-306-44690-0.
- Halmos, Paul (1963). Lectures on Boolean Algebras. van Nostrand.
- Bacon, Jason W. (2011). "Computer Science 315 Lecture Notes". Retrieved October 1, 2021.
- O'Regan, Gerard (2008). A brief history of computing. Springer. p. 33. ISBN 978-1-84800-083-4.
- Atiyah, Michael Francis; Macdonald, I. G. (1969), Introduction to Commutative Algebra, Westview Press, ISBN 978-0-201-40751-8
- Fraleigh, John B. (1976), A First Course In Abstract Algebra (2nd ed.), Addison-Wesley, ISBN 978-0-201-01984-1
- Herstein, I. N. (1975), Topics In Algebra (2nd ed.), John Wiley & Sons
- McCoy, Neal H. (1968), Introduction To Modern Algebra (Revised ed.), Allyn and Bacon, LCCN 68015225
- Ryabukhin, Yu. M. (2001) [1994], "Boolean ring", Encyclopedia of Mathematics, EMS Press
- "Elements of Boolean Algebra". www.ee.surrey.ac.uk. Retrieved 2020-09-02.
- Peter J. Pahl; Rudolf Damrath (2012-12-06). Mathematical Foundations of Computational Engineering: A Handbook. Springer Science & Business Media. pp. 15–. ISBN 978-3-642-56893-0.
- Hall, Eldon C. (1996). Journey to the Moon: The History of the Apollo Guidance Computer. AIAA. ISBN 1-56347-185-X.
- "APOLLO GUIDANCE COMPUTER (AGC) Schematics". klabs.org.

Rich Katz. Retrieved 2021-06-19. To see how NOR gate logic was used in the Apollo Guidance Computer's ALU, select any of the 4-BIT MODULE entries in the Index to Drawings, and expand images as desired.

- Bender, Edward A.; Williamson, S. Gill (2005). A Short Course in Discrete Mathematics. Mineola, NY: Dover Publications, Inc. ISBN 0-486-43946-1.
- McCluskey, E. J. (1965). Introduction to the Theory of Switching Circuits. NY: McGraw–Hill Book Company. p. 78. LCCN 65-17394. Canonical expressions are defined and described
- Hill, Fredrick J.; Peterson, Gerald R. (1974). Introduction to Switching Theory and Logical Design (2nd ed.). NY: John Wiley & Sons. p. 101. ISBN 0-471-39882-9. Minterm and maxterm designation of functions
- For bitwise operations in computer programming, it may be helpful to read 1 as 0xFFFF. All bits of the binary number must be 1.
- *Goodstein, R. L. (2012), "Chapter 4: Sentence Logic", Boolean Algebra, Courier Dover Publications, ISBN 9780486154978
- Steven R. Givant; Paul Richard Halmos (2009). Introduction to Boolean algebras. Springer. pp. 21–22. ISBN 978-0-387-40293-2.
- Venn, John (July 1880). "I. On the Diagrammatic and Mechanical Representation of Propositions and Reasonings" (PDF). The London, Edinburgh, and Dublin Philosophical Magazine and Journal of Science. 5. 10 (59): 1–18. doi:10.1080/14786448008626877. Archived (PDF) from the original on 2017-05-16.
- Shannon, Claude (1949). "The Synthesis of Two-Terminal Switching Circuits". Bell System Technical Journal. 28: 59–98. doi:10.1002/j.1538-7305.1949.tb03624.x.
- Stavi, Jonathan (1974), "A model of ZF with an infinite free complete Boolean algebra", Israel Journal of Mathematics, 20 (2): 149–163, doi:10.1007/BF02757883, S2CID 119543439.
- Johnstone, Peter T. (1982), Stone spaces, Cambridge University Press, ISBN 0-521-33779-8
- Koppelberg, Sabine (1989), Monk, J. Donald; Bonnet, Robert (eds.), Handbook of Boolean algebras, vol. 1, Amsterdam: North-Holland Publishing Co., pp. xx+312, ISBN 0-444-70261-X, MR 0991565
- Monk, J. Donald; Bonnet, Robert, eds. (1989), Handbook of Boolean algebras, vol. 2, Amsterdam: North-Holland Publishing Co., ISBN 0-444-87152-7, MR 0991595
- Monk, J. Donald; Bonnet, Robert, eds. (1989), Handbook of Boolean algebras, vol. 3, Amsterdam: North-Holland Publishing Co., ISBN 0-444-87153-5, MR 0991607

- Vladimirov, D.A. (2001) [1994], "Boolean algebra", Encyclopedia of Mathematics, EMS Press
- Koppelberg, Sabine (1989). "General Theory of Boolean Algebras". Handbook of Boolean Algebras, Vol. 1 (ed. J. Donald Monk with Robert Bonnet). Amsterdam: North Holland. ISBN 978-0-444-70261-6.
- McCune, William; Veroff, Robert; Fitelson, Branden; Harris, Kenneth; Feist, Andrew; Wos, Larry (2002), "Short single axioms for Boolean algebra", Journal of Automated Reasoning, 29 (1): 1–16, doi:10.1023/A:1020542009983, MR 1940227, S2CID 207582048
- Allwood, Jens; Andersson, Gunnar-Gunnar; Andersson, Lars-Gunnar; Dahl, Osten (1977-09-15). Logic in Linguistics. Cambridge University Press. ISBN 978-0-521-29174-3.
- Hausman, Alan; Howard Kahane; Paul Tidman (2010) [2007]. Logic and Philosophy: A Modern Introduction. Wadsworth Cengage Learning. ISBN 978-0-495-60158-6.
- Girard, Jean-Yves; Paul Taylor; Yves Lafont (1990) [1989]. Proofs and Types. Cambridge University Press (Cambridge Tracts in Theoretical Computer Science, 7). ISBN 978-0-521-37181-0.
- Not all search engines support the same query syntax. Additionally, some organizations (such as Google) provide "specialize≤ search engines that support alternate or extended syntax. (See e.g.,Syntax cheatsheet, Google codesearch supports regular expressions).

INDEX

9 789390 013609

Printed at Repro India Ltd.